D1314124

A PRIMER ON
Sediment-Trace Element Chemistry

SECOND EDITION

Arthur J. Horowitz

LEWIS PUBLISHERS

Library of Congress Cataloging-in-Publication Data

Horowitz, Arthur J.
    A primer on sediment-trace element chemistry / by Arthur J.
Horowitz. -- 2nd rev. ed.
        p.   cm.
    Rev. ed. of: A primer on trace metal-sediment chemistry. 1985.
    Includes bibliographical references and index.

    1. Trace elements. 2. Metals. 3. Geochemistry. 4. Sediments
(Geology)   I. Horowitz, Arthur J. Primer on trace metal-sediment
chemistry. II. Title.
QE516.T85H67   1991
551.9--dc20   90-23488
ISBN 0-87371-499-7

LEWIS PUBLISHERS, INC.
121 South Main Street, Chelsea, Michigan 48118

PRINTED IN THE UNITED STATES OF AMERICA

For Lyn and William, who had to put up with my endless 'progress reports', and for my father, who dreamed, but didn't see it come to fruition.

ACKNOWLEDGMENTS

I would like to thank the U.S. Geological Survey's Water Resources Division, and particularly, the Office of Water Quality, for their support over the past seven years. Without it, much of my own research, as well as the production of this second revised edition, would not have been possible. I particularly want to thank Drs. Wes Bradford and R. J. Pickering (who also suggested the term 'Primer') because without their initial encouragement and active support, much of the ensuing work would have been impossible. I also would like to thank Charles Demas, Bernard Malo, James Miller, and Gary Pederson for their critical reviews and suggestions for this second revised edition. Finally, I would like to thank all the students who have sat through my lectures on sediment-trace element chemistry, particularly those who were sufficiently stimulated to ask questions. Those questions continued to show the truth of Lord Kelvin's oft-cited statement that: 'If you can't explain it, you don't understand it'! They caused me to reexamine and rethink much of what I have taken as 'givens' in the field of sediment-trace element chemistry.

PREFACE

The first edition of this Primer originally was developed in 1983, and published in 1985 as a U.S. Geological Survey Water Supply Paper 2277. It was designed to serve as a companion training volume for a U.S. Geological Survey short course on sediment-trace element chemistry. The purpose of the course was to expose the participants to the basic principles and techniques that control and are used to elucidate sediment-trace element interactions. Participants were not expected to walk away from the course as experts in sediment chemistry; however, they were expected to have a good working knowledge of the basic principles governing this complex subject.

The first edition was organized in the same way as the instructional sessions, contained all the diagrams and tables used for the course, along with a descriptive text, and was designed to be self-teaching. To that end, it was produced using a 'stop' format with text on one page and supporting material, diagrams, and/or tables appearing on the facing page. It also included a large list of selected references on the subject of sediment-trace element chemistry. This list was by no means exhaustive. Many of the references were used in developing the material in the course and in the Primer. The other references were provided for information on techniques and methods, and as examples of how sediment chemistry could be used to deal with various types of environmental studies.

It has been more than seven years since the preparation and publication of the first edition. The basic course has been taught more than ten times and, in addition, a more comprehensive team-taught solid-phase course has been introduced. Further, as in any active scientific field, much research has been carried out and, as a result, our understanding of this complex subject has increased commensurately. These factors have led to the production of this second revised edition.

CONTENTS

Glossary

Bioavailability — In general, refers to that portion of dissolved, biologically-, or sediment-associated chemical constituents that are readily accessible to biota either through physical contact or by ingestion; this is an operationally defined term.

Bottom sediment, bed sediment, bed material — Interchangeable terms that refer to material that temporarily is stationary in the bottom of a water course.

Dissolved load, dissolved constituent — Operationally defined as that material that passes through a 0.45-μm (micrometer) filter.

Geochemical factors — Refers to various physical and chemical characteristics which can affect a sediment's chemical concentration.

Geochemical substrate — Refers to the compounds or substances that are important in providing a sediment with the capacity to collect and retain trace elements.

Isokinetic — A term used to describe a type of suspended sediment sampling in which a whole water sample is collected such that no acceleration or deceleration is imparted to the mixture as it enters the sampling device.

Operationally defined — Refers to the quantitation of a physical or chemical constituent or property which is dependent on the method used for its determination. If the method is altered the measurement changes also; many of the physical and chemical determinations performed on sediment/solid samples fall into this category.

Sediment — Particles derived from rocks or biological materials that have been transported by a fluid, or solid material suspended in or settled from water.

Suspended load, suspended constituent — Operationally defined as that material that is retained by a 0.45-μm (micrometer) filter during the filtration of a whole water sample.

Suspended sediment, seston, particulates — Interchangeable terms that refer to a material actively transported by a fluid.

Total concentration —Refers to the true concentration of a constituent in a sediment/solid sample; geochemists further define total as ≥95 percent of the actual concentration in the material being analyzed. Certain analytical procedures, or digestion techniques can produce a measure of total concentration without the complete dissolution of a solid sample. Unless otherwise stated, concentrations in this Primer are totals.

Total recoverable concentration — Refers to the concentration of a constituent in a sediment/solid sample that has been extracted with a reagent(s) that does not completely solubilize the constituent; in other words, this is an operationally defined concentration, the definition being dependent upon the extraction procedure.

Trace element, trace metal, heavy metal, metal — Terms that are used interchangeably throughout the sediment-chemical literature and this Primer.

INTRODUCTION

The basic goals of most chemically oriented water-quality studies are to describe or evaluate existing environmental conditions and to attempt to identify the source or sources of the constituents under investigation. An equally important goal is to attempt to predict or determine potential effects. This heading could accommodate such subjects as bioavailability, amount of constituent transport, location of chemical sinks, ultimate fate, and potential toxic effects.

Historically, most water quality investigations have attempted to assess trace elements in aquatic systems by analyzing water samples. This assessment has entailed determining concentrations of total and dissolved elements and compounds through the collection and analysis, respectively, of unfiltered and filtered water. Concentrations associated with suspended sediment (particulates, seston) are determined indirectly by the difference between total and dissolved concentrations. It is recognized that this approach casts doubt on the reliability of reported suspended-sediment chemical analyses. As a result, water quality tends to be evaluated on the kinds and concentrations of various constituents found in solution (Feltz, 1980). Further, current federal, state, and local regulations usually only provide limits for dissolved trace element concentrations; concerns about regulations tend to be the cause for many studies. However, in most aquatic systems, the concentrations of trace elements in suspended sediment and the top few centimeters of bottom sediment are far greater (by orders of magnitude) than the concentration of trace elements in the water column. The strong association of numerous trace elements (e.g., As, Cd, Hg, Pb, Zn) with seston and bottom sediments means that the distribution, transport, and availability of these constituents can not be evaluated intelligently solely through the sampling and analysis of the dissolved phase.

Additionally, because bottom sediments can act as a reservoir for many trace elements, they must, for several reasons, be given serious consideration in the planning and design of any water-quality study. First, an undisturbed sediment sink contains a historical record of chemical conditions. If a sufficiently large and stable sink can be located and studied, it will allow an investigator to evaluate chemical changes over time and, possibly, to establish area baseline levels against which current conditions can be compared and contrasted. Second, under changing environmental or physicochemical conditions (e.g., pH, Eh, dissolved oxygen, bacterial action), sediment-bound trace elements can dissolve into the water column, possibly enter the food chain, and have significant environmental effects. Third, several relatively inert or otherwise environmentally harmless inorganic constituents can degrade, or react with others, to form soluble and potentially more toxic forms (e.g., the conversion of elemental mercury to methyl mercury, the conversion of arsenopyrite to iron oxide-associated arsenic). Fourth, bottom sediments should be regarded as a major source of suspended sediment. Therefore, they must be investigated to determine transport potential. Under changing hydrologic conditions (such as a heavy storm or spring runoff), a localized pollution problem suddenly can become widespread and result in substantial environmental impacts. Finally, bottom sediment sampling and subsequent chemical analyses can provide fundamental regional geochemical information for use in identifying potential environmental effects due to both point and non-point sources of contamination.

The foregoing discussion indicates that data on suspended and bottom sediments, as well as on the dissolved phase, are a requisite for the development of a comprehensive understanding of the effects of trace elements on water quality. Through the use of such additional data, it may be possible to begin to identify sources and sinks and the fate and potential effects of toxic or environmentally necessary trace elements. Similarly, sediment-chemical data are a requisite for transport modelling, for estimating geochemical cycles, and for inferring the availability of various trace elements in and to an ecological system.

1.0 Importance of Sediments to Aquatic Trace Element Chemistry

1.1 Monitoring Studies

Table 1.1-1 has been extracted from an article by Chapman et al. (1982) dealing with the design of monitoring studies for priority pollutants. They divide the aquatic environment into three distinct compartments: water, sediment, and biota. Further, the authors point out that, to design an effective monitoring program, an investigator must first answer two basic questions: what samples should be collected, and for what should they be analyzed? This decision requires an understanding of the relative importance of each constituent on the basis of its chemical behavior and biological significance. These factors, in turn, determine what type of sample(s) should be collected (water, sediment, biota). All the U.S. Environmental Protection Agency trace element priority pollutants, with the exception of antimony, are persistent and can bioaccumulate, and all tend to be concentrated in either sediment (bottom and suspended) or biota (Table 1.1-1).

1.0 Importance of Sediments to Aquatic Trace Element Chemistry

1.2 Suspended versus bed sediments - utility for various types of studies

Table 1.2-1 lists a number of different types of environmental/water quality studies in which sediment sampling and subsequent chemical analysis could and should be included. In almost all cases, with the possible exception of studies designed to evaluate the transport of (including the calculation of trace element fluxes) or short-term temporal variations in sediment-associated trace elements, either suspended or bed sediments could be used. However, given an option, bed sediments tend to be preferable to suspended sediments. There are two major reasons for this selection: 1) when suspended sediment concentrations are relatively low (<100 mg/L) it is usually far simpler and easier to collect sufficient amounts of bed sediment than suspended sediment to meet the mass requirements for all requisite physical and chemical analyses, and 2) suspended sediments tend to display much more marked spatial and temporal chemical and physical variability than do bed sediments (Forstner and Wittmann, 1981; Salomons and Forstner, 1984; Ongley, et al., 1988; Horowitz, et al., 1989c, 1990). Although there are several ways to collect and process sufficient amounts of suspended sediment to meet most analytical requirements, the procedures involved and/or the equipment necessary, are either very time consuming and/or very expensive (Carpenter, et al., 1975; Etchebar and Jouanneau, 1980; Ongley and Blachford, 1982; Horowitz, 1986; Horowitz, et al., 1989b).

Table 1.1-1. Recommendations for Types of Environmental Sampling for Monitoring Purposes (Data from Chapman, et al., 1982)

Trace Element	Category/Rank*	Water	Sediment	Biota
Antimony (Sb)	3	X		
Arsenic (As)	1		X	X
Beryllium (Be)	1		X	X
Cadmium (Cd)	1		X	X
Chromium (Cr)	1		X	X
Copper (Cu)	1		X	X
Lead (Pb)	1		X	X
Mercury (Hg)	1		X	X
Selenium (Se)	1		X	X
Silver (Ag)	1		X	X
Thallium (Tl)	1		X	X
Zinc (Zn)	1		X	X

Category/Rank*: 1 and 2 - persistent and may bioaccumulate

3 and 4 - persistent and nonaccumulative

5 - nonpersistent

Table 1.2-1. Various Types of Trace Element Studies in Which Sediment Sampling and Analysis Should be Included

Bed Sediment	Type of Study	Suspended Sediment
X	Geochemical Cycles	X
	Transport	X
	Fluxes	X
X	Reconnaissance Surveys	X
X	Spatial Distributions	X
X	Temporal Changes (Long Term)	X
	Temporal Changes (Short Term)	X
X	Biological Effects	X

3

1.0 IMPORTANCE OF SEDIMENTS TO AQUATIC TRACE ELEMENT CHEMISTRY

1.3 Partitioning of trace elements between dissolved and solid phases

As stated previously, both bottom and suspended sediments contain significantly higher concentrations of trace elements than are found in the dissolved phase. Hence, fluvial transport of these constituents can be dominated by suspended sediment. Figure 1.3-1 is based upon the analysis of dissolved and suspended loads for selected trace elements in the Amazon (Brazil) and Yukon Rivers (Alaska) and graphically displays this phenomenon (adapted from Gibbs, 1977). In addition, Martin and Meybeck (1979) and Meybeck and Helmer (1989) have looked at the ratio of natural dissolved to total elemental fluvial transport of a substantially larger number of trace elements and rivers (Table 1.3-1). Sb appears to be the most soluble of all the priority trace element pollutants (roughly 50 percent of Sb transport is associated with solid phases) while Hg appears to be the least soluble (>99.9 percent of Hg transport is associated with solid phases). The other trace elements fall somewhere in between, but for the great majority, 90 percent or more of their transport is associated with solid phases. This statement is an oversimplification and somewhat misleading because trace element transport in aquatic systems is dependent on several factors including: discharge, dissolved trace element concentrations, suspended sediment-associated trace element concentrations, as well as the concentration of suspended sediment (e.g., see Section 1.6).

4

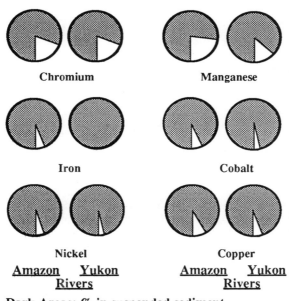

Figure 1.3-1. Transport Phases of Selected Trace Elements in Rivers (Data from Gibbs, 1977)

Chromium   Manganese

Iron   Cobalt

Nickel   Copper

Amazon   Yukon   Amazon   Yukon
      Rivers   Rivers

Dark Areas: % in suspended sediment
Light Areas: % in solution

Table 1.3-1. Ratio Between Natural Dissolved and Total Elemental River Transport[1] (from Meybeck and Helmer, 1989)

99% 90%	50%	10% 5%
Cl│ Br S │ Na Sr C Ca Li│ Sb Mg N B Mo As F Ba K│ Cu P│ Ni Si Rb U Co Cd
Mn Th V Cs│ Ga Pb Lu│ Ti Gd La Ho Yb Tb Er Sm Cr Fe Eu Ce Zn Al│ Sc Hg
1%	0.5%	0.1% 0.05%

[1]The higher the percentage, the greater the partitioning into the dissolved phase

1.0 IMPORTANCE OF SEDIMENTS TO AQUATIC TRACE ELEMENT CHEMISTRY

1.4 Fluvial transport of trace elements by suspended sediments

As a further indication that fluvial transport of trace elements may be dominated by the suspended-sediment load, it is worthwhile to determine, in relatively simple terms, what order of magnitude of material is being transported. Examine the data in Table 1.4-1. The discharge equation comes from Porterfield (1972), and the basic data for the actual calculations come from U.S. Geological Survey annual data reports from Pennsylvania (Susquehanna River), Oregon (Willamette River), and Louisiana (Mississippi River). Note that even relatively low concentrations of trace elements associated with suspended sediment can involve significant weights of constituent transport. For example, the As concentration in Willamette River suspended sediment is only 0.002 mg/L (2 ppb), and yet this translates into 0.9 metric tons/day of As being transported by the river. At the other end of the spectrum, examine the Fe data for the Mississippi River. The suspended-sediment Fe concentration is 46 mg/kg (46 ppm); thus some 75,000 metric tons/day of Fe are transported. As the chemical concentrations for the suspended sediment were determined on whole water samples, using a total recoverable procedure, the values for daily transport are almost certainly underestimates relative to total trace element transport.

Table 1.4-1. Transport (in metric tons/day) of Trace Elements Associated with Suspended Sediment in Selected Rivers

Element	Susquehanna River 3/8/79		Willamette River 1/16/80		Mississippi River 1/18/81	
Discharge[1]	346,000 CFS		189,000 CFS		671,000 CFS	
	Concentration[2] in mg/L	Transport[3] in metric tons/day	Concentration[2] in mg/L	Transport[3] in metric tons/day	Concentration[2] in mg/L	Transport[3] in metric tons/day
Sediment	406	344,200	106	49,100	641	1,050,000
Iron	12	10,200	4.4	2,040	46	75,600
Manganese	0.59	500	0.10	46	1.3	2,140
Zinc	0.07	59	0.03	14	0.09	148
Lead	0.018	15	0.005	2.3	0.03	49
Copper	0.015	13	0.006	2.8	0.018	30
Cobalt	0.010	8.5	-	-	0.005	8.2
Chromium	0.003	2.5	0.010	4.6	0.028	46
Nickel	-	-	0.004	1.9	0.035	58
Arsenic	-	-	0.002	0.9	0.012	20

[1]Discharge - in cubic feet per second
[2]Concentration - total recoverable, in milligrams per liter
[3]Sediment and chemical transport in metric tons per day, calculated from Porterfield, 1972:
    (concentration, in milligrams per liter)(discharge, in cubic feet per second) 0.00245

1.0 IMPORTANCE OF SEDIMENTS TO AQUATIC TRACE ELEMENT CHEMISTRY

1.5 Comparison of trace element concentrations in suspended and bottom sediments versus dissolved levels

Bottom and suspended sediments contain substantially higher concentrations of trace elements than are found in solution. What is meant by substantially higher? Examination of the data presented in the accompanying tables provides an approximate answer to this question. Table 1.5-1 contains data from the Elbe River in the Federal Republic of Germany (data from Forstner and Wittmann, 1981), and compares and contrasts dissolved trace element concentrations with those associated with bottom sediments. The bottom sediment concentrations are more than 100,000 (5 orders of magnitude) times higher than the dissolved levels (see, for example, Pb levels for water and sediment for Hamburg). During the Schuylkill River Basin assessment study, similar results were obtained for dissolved and total recoverable levels (Table 1.5-2, from Feltz, 1980). Table 1.5-3 shows data from the Amazon and Yukon Rivers and compares and contrasts dissolved trace element concentrations with those associated with suspended sediment. The suspended sediment concentrations for the Amazon and Yukon are approximately 10,000 and 7,000 times higher, respectively, than the dissolved load.

Table 1.5-1. Comparison of Trace Element Concentrations in River Water and Bottom Sediments from the Elbe River, FRG (Data from Forstner and Wittmann, 1981)

Location	Type	Cd	Cr	Cu	Pb	Zn
Stade	(W)[*]	.005	.003	.005	.005	.025
	(S)[*]	5	100	100	200	600
Hamburg	(W)	.0007	.010	.012	.005	.120
	(S)	32	500	450	500	1600
Hitzacker	(W)	.0008	.020	.020	.008	.140
	(S)	23	250	200	200	750

(W)[*] - water, trace element concentrations in milligrams per liter
(S)[*] - sediment, trace element concentrations in milligrams per kilogram

Table 1.5-2. Comparison of Trace Element Concentrations in River Water and Bottom Sediments from the Schuylkill Basin (Data from Feltz, 1980)

Location	Type	Cd	Cr	Cu	Pb	Zn
Phoenixville	(W)[*]	.013	.001	.005	.190	.080
	(S)[*]	30	130	190	250	1000
Norristown	(W)	.010	< .001	.007	.110	.120
	(S)	10	30	70	400	1000
Philadelphia	(W)	.002	< .001	.004	.001	< .001
	(S)	10	50	80	180	170

(W)[*] - water, trace element concentrations in milligrams per liter
(S)[*] - sediment, trace element concentrations in milligrams per kilogram

Table 1.5-3. Comparison of Trace Element Concentrations in River Water and Suspended Sediment from the Amazon and Yukon Rivers (Data from Gibbs, 1977)

Location	Type	Cr	Mn	Fe	Co	Ni	Cu
Amazon	(W)[*]	.019	.110	5.1	.004	.010	.026
	(S)[*]	195	1100	56,000	41.4	105	265
Yukon	(W)	.019	.200	8.9	.006	.020	.060
	(S)	115	1300	63,000	40.6	135	415

(W)[*] - water, trace element concentrations in milligrams per liter
(S)[*] - sediment, trace element concentrations in milligrams per kilogram

1.0 IMPORTANCE OF SEDIMENTS TO AQUATIC TRACE ELEMENT CHEMISTRY

1.6 Effect of suspended sediment concentration on fluvial transport of trace elements

The partitioning (percentage association) of various trace elements between liquid and solid phases, during transport, is a function of three factors: 1) the concentration of suspended sediment (mg/L, ppm), 2) the dissolved trace element concentration (μg/L, ppb), and 3) the sediment-associated trace element concentration (μg/g, mg/kg, ppm). Although suspended sediment-associated trace element concentrations are orders of magnitude higher than dissolved trace element concentrations, there are orders of magnitude more water than sediment. Thus, for purposes of transport, the contributions from the substantially higher concentrations of sediment-associated trace elements may be balanced by the substantially lower concentrations of suspended sediment, relative to the amount of water, in typical fluvial systems. Table 1.6-1 contains data on trace element partitioning for a set of hypothetical whole water samples which substantiates this statement. The dissolved data come from a selection of unpolluted rivers around the world (Forstner and Wittmann, 1981) and the sediment data comes from a selection of bed sediment samples from unpolluted rivers in the U.S. (Horowitz, et al, 1989a). The data clearly show that as suspended sediment concentrations increase, the percentage contribution from suspended sediment-associated trace elements also increases. In the case of such elements as Pb, Zn, Cr, Co, and Cu, solid phase contributions are greater than 50 percent, even when suspended sediment concentrations are as low as 10 mg/L. On the other hand, suspended sediment concentrations must be 100 mg/L or higher, before the solid phase contributions for such elements as Se, Hg, As, and Sb begin to approach or exceed 50 percent (Table 1.6-1). These results indicate why the data in Table 1.3-1 from Meybeck and Helmer (1989) can be misleading regarding the importance of sediment-associated concentrations, relative to dissolved concentrations, for the fluvial transport of trace elements.

Table 1.6-1. Solid Phase Contributions to Riverine Transport of Trace Elements for Various Suspended Sediment Concentrations [Dissolved trace element data from Forstner and Wittmann (1981); solid phase trace element data from Horowitz, et al. (1989a)]

Element	Dissolved Conc.[2] µg/L	Solid Conc.[3] µg/L	10 mg/L[1] Solid Cont.[4] µg/L	10 mg/L[1] Total Conc.[5] µg/L	10 mg/L[1] Solid Cont.[6] %	100 mg/L[1] Solid Cont. µg/L	100 mg/L[1] Total Conc. µg/L	100 mg/L[1] Solid Cont. %	500 mg/L[1] Solid Cont. µg/L	500 mg/L[1] Total Conc. µg/L	500 mg/L[1] Solid Cont. %	1000 mg/L[1] Solid Cont. µg/L	1000 mg/L[1] Total Conc. µg/l	1000 mg/L[1] Solid Cont. %
Cu	0.2	25	0.25	0.45	56	2.5	2.7	93	12.5	12.7	98	25.0	25.2	>99
Zn	0.2	90	0.90	1.10	82	9.0	9.2	98	45	45.2	>99	90.0	90.2	>99
Cd	0.01	0.6	0.006	0.016	38	0.06	0.07	86	0.3	0.31	97	0.6	0.61	>99
Cr	0.07	20	0.2	0.27	74	2.0	2.07	97	10.0	10.07	>99	20.0	20.07	>99
Pb	0.05	50	0.5	0.55	91	5.0	5.05	99	25.0	25.05	>99	50.0	50.05	>99
Co	0.05	18	0.18	0.23	78	1.8	1.85	97	9.0	9.05	>99	18.0	18.05	>99
Ni	0.3	25	0.25	0.55	46	2.5	2.8	89	12.5	12.8	98	25.0	25.3	>99
As	0.5	7	0.07	0.57	12	0.7	1.2	58	3.5	4.0	88	7.0	7.5	93
Sb	0.05	0.6	0.006	0.056	11	0.06	0.11	55	0.3	0.35	86	0.6	0.65	92
Se	0.08	0.4	0.004	0.084	5	0.04	0.12	33	0.2	0.28	71	0.4	0.48	83
Hg	0.006	0.05	0.0005	0.0065	8	0.005	0.011	46	0.025	0.031	81	0.05	0.056	89

[1]Suspended sediment concentration

[2]Dissolved Conc. - dissolved trace element concentration

[3]Solid Conc. - suspended sediment-associated trace element concentration

[4]Solid Cont. - solid contribution, see formula below

[5]Total Conc. - total trace element concentration for the whole water sample, see formula below

[6]Solid Cont. - solid contribution, calculated by dividing the solid contribution by the total sample concentration and multiplying by 100

Solid Trace Element Cont. (µg/L) = [Trace Element S.S. Conc. (µg/g)] [S.S. Conc. (g/L)]
Total Conc. (µg/L) = [Solid Trace Element Cont. (µg/L)] + [Dissolved Conc. (µg/L)]

1.0 IMPORTANCE OF SEDIMENTS TO AQUATIC TRACE ELEMENT CHEMISTRY

1.7 Historical trace element levels

An undisturbed sediment sink can contain a historical record of chemical conditions. If a sufficiently large and stable sink (one in which physical, biological, nor chemical alterations have occurred) can be found and studied, an investigator can establish the occurrence of chemical changes over time and may establish area baseline levels with which existing conditions can be compared. When chemical analyses are used together with radiometric dating techniques (^{210}Pb, ^{14}C), historical changes in water quality can be elucidated. Figure 1.7-1 gives two examples of historical records detailing changes in Hg concentration through time. The data from Lake Ontario (Thomas, 1972) indicate that high concentrations of Hg occur in the top 6 cm; below 8 cm, the concentration decreases significantly. The levels found below 30 cm are believed to reflect natural background levels of 0.3 to 0.4 mg/kg (300 to 400 ppb). At about 25-cm in depth, the concentration reaches a low of 0.140 mg/kg (140 ppb), which corresponds to a period of active deforestation by settlers around 1800 to 1820. That activity, in fact, diluted the Hg levels below normal background for the area through increased erosion and deposition of non-Hg-bearing material. Industrial input is believed to have begun around 1900 (~9 cm); there ensued a steady rise in Hg input until around 1940 (~5 cm), when concentrations leveled off at about 4 times natural levels.

A similar pattern can be seen in sediments from Lake Windermere in England (Aston, et al., 1973). Since 1400, Hg levels have risen steadily as a result of erosion, industrialization, mining, quarrying, use of fossil fuels, and sewage disposal. The onset of industrial input occurred around 1880 (~24 cm), and today's levels are about 4 to 5 times higher than natural background concentrations. It is interesting to note that this is roughly the same level of enhancement as seen in Lake Ontario sediments.

Figure 1.7-1 Examples of Mercury Concentration Changes Through Time [Lake Ontario, Canada data from Thomas (1972); Lake Windermere, United Kingdom data from Aston et al. (1973)]

Mercury in Lake Bed Sediments in Milligrams per Kilogram

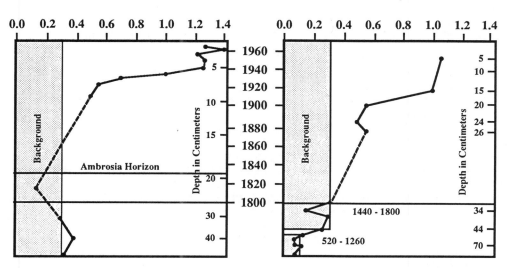

Lake Ontario, Canada

Lake Windermere, England

2.0 PHYSICAL AND CHEMICAL FACTORS AFFECTING SEDIMENT-TRACE ELEMENT CHEMISTRY

2.1 Introduction

The preceding material is based on the determination of total bottom or suspended sediment chemical concentrations. These types of chemical data are typical and standard starting points for most water-quality studies. However, a number of sedimentary physical and chemical measurements (geochemical factors) should be considered as a requisite for understanding sediment-trace element chemistry. These measurements enable an investigator to address such topics as trace element transport and potential environmental effects, and permit the comparison of sample data from the same or different areas. These factors describe and/or are measurements of the various chemical or compositional components and physical properties of a sediment.

The identification and quantitation of various sedimentary geochemical factors is sometimes called partitioning. *Physical partitioning* refers to the separation of a sediment sample (bottom or suspended) into various categories (by grain size, surface area, specific gravity, magnetic properties, etc.,Table 2.1-1). *Chemical partitioning* refers to the separation, identification, and quantitation of the various chemical components (sometimes called geochemical substrates) found in sediments as well as to the type(s) of association various trace elements have with these geochemical substrates. Chemical partitioning is further subdivided into a chemical phase group which deals with the identification and quantitation of various geochemical substrates such as carbonates, clay minerals, organic matter, iron and manganese oxides and hydroxides, silicates, etc. and into a chemical mechanistic group which deals with the type(s) of association that exist between trace elements and various geochemical substrates such as adsorption, precipitation, organometallic bonding, incorporation in crystal lattices, etc. (Table 2.1-1).

Although the two separate definitions imply that the physical and chemical factors are independent of each other, they are, in fact, strongly interrelated (Fig. 2.1-1). As such, the separation into physical and chemical categories is somewhat arbitrary. For convenience, the two types of partitioning initially are discussed separately (Section 2.2 for the physical factors and Section 2.3 for the chemical factors); the inter- and intrarelations between and among the various physical and chemical factors will be discussed later (see Section 2.4).

14

Table 2.1-1. Examples of Types of Physical and Chemical Partitioning

Physical	Chemical Mechanistic	Chemical Phase
Grain Size	Adsorption	Interstitial Water
Surface Area	Precipitation	Carbonates
Specific Gravity	Co-Precipitation	Clay Minerals
Surface Charge	Organometallic Bonding	Hydrous Fe and Mn
	Cation Exchange	Oxides
	Incorporation in Crystalline	Sulfides
	Minerals	Silicates

Figure 2.1-1. Geochemical Factors Affecting Sediment-Trace Element Chemistry

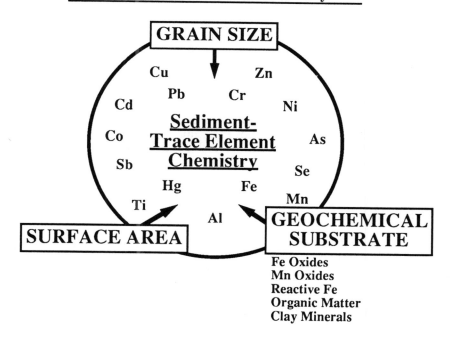

15

2.0 PHYSICAL AND CHEMICAL FACTORS AFFECTING SEDIMENT-TRACE ELEMENT CHEMISTRY

2.2 *Physical factors*

Sediments can be analyzed for a wide variety of physical attributes (Table 2.2-1). However, from both a water quality and a geochemical perspective, only two physical factors require substantial evaluation. These are grain size and surface area.

2.0 PHYSICAL AND CHEMICAL FACTORS AFFECTING SEDIMENT-TRACE ELEMENT CHEMISTRY

2.2 *Physical factors*

2.2.1 Grain-size ranges and the effect of grain size

One of the most significant factors controlling both suspended- and bottom-sediment capacity for concentrating and retaining trace elements is grain size (Goldberg, 1954; Krauskopf, 1956; Goldberg and Arrhenius, 1958; Hirst, 1962; Jenne, 1968; Kharkar, et al., 1968; Gibbs, 1977; Jones and Bowser, 1978; Filipek and Owen, 1979; Jenne, et al., 1980; Thorne and Nickless, 1981; Horowitz and Elrick, 1987). Table 2.2.1-1 lists the sediment particle-size classes (names) and the size ranges the names represent. In lakes, rivers, estuaries, and oceans—and in sediment chemistry in general—most sediments tend to be composed of materials smaller (finer) than 2,000 μm (very coarse sand).

There is a very strong positive correlation between decreasing grain size and increasing trace element concentrations. This correlation results from a combination of both physical (e.g., surface area) and chemical factors (e.g., geochemical substrates). Clay-sized sediments (<2 to 4 μm, Table 2.2.1-1) have surface areas measured in square meters per gram as opposed, for example, to sand-sized particles with surface areas commonly measured in tens of square centimeters per gram (Grim, 1968; Jones and Bowser, 1978). Surface chemical reactions are extremely important to aquatic trace element-sediment interactions; thus, fine-grained sediments, because of their large surface areas, are the main sites for the collection and transport of inorganic constituents (Krauskopf, 1956; Jenne, 1968; Gibbs, 1973; Jones and Bowser, 1978; Jenne, et al., 1980). However, Jenne (1976) indicates that clay-sized particles may be viewed simply as mechanical substrates upon which trace elements can concentrate (without chemical interaction). Bear in mind that trace element concentrations can and do accumulate on coarse material as well, including sand, pebbles, cobbles, and boulders (Filipek, et al., 1981; Robinson, 1982); nevertheless, high concentrations are more commonly associated with fine-grained material.

A number of other factors also have substantial effects on trace element concentrations (see Table 2.1-1); however, there is a 'common thread' for all these factors: *grain size*. As grain size decreases, these additional factors become more important, and a positive correlation usually exists between them and increasing trace element concentration (Jones and Bowser, 1978; Forstner and Wittmann, 1981; Salomons and Forstner, 1984; Horowitz and Elrick, 1987). In some instances, it is impossible to differentiate between effects caused by factors like surface area, cation exchange capacity, surface charge, and the increasing concentration of various geochemical substrates and effects due to grain size. Hence, if only one physical property is to be determined to aid in interpreting chemical data, grain size is by far the property of choice, because it seems to integrate all the others.

Table 2.2-1. Examples of Types of Sedimentary Physical Measurements

Grain Size
Surface Area
Specific Gravity
Surface Charge
Bulk Density
Sheer Stress
Porosity
Permeability

Table 2.2.1-1. Sediment Particle Sizes and Class Names (modified Wentworth - Udden Scale)

<u>Class Name</u>	<u>Millimeters</u>	<u>Micrometers</u>	
Boulders	>256		
Large Cobbles	256 - 128		
Small Cobbles	128 - 64		
V. Coarse Gravel	64 - 32		
Coarse Gravel	32 - 16		
Medium Gravel	16 - 8		
Fine Gravel	8 - 4		
V. Fine Gravel	4 - 2		
V. Coarse Sand		2000 - 1000	
Coarse Sand		1000 - 500	
Medium Sand		500 - 250	
Fine Sand		250 - 125	
V. Fine Sand		125 - 62.5	
Coarse Silt		62.5 - 31	
Medium Silt		31 - 16	
Fine Silt		16 - 8	
V. Fine Silt		8 - 4	
Coarse Clay		4 - 2[*]	
Medium Clay		2 - 1	
Fine Clay		1 - 0.5	
V. Fine Clay		0.5 - 0.25	

2[*] - many sedimentologists consider that the silt/clay break occurs at 2μm rather than at 4μm.

2.0 PHYSICAL AND CHEMICAL FACTORS AFFECTING SEDIMENT-TRACE ELEMENT CHEMISTRY

2.2 Physical factors

2.2.2 Chemical analysis of various grain sizes in bottom sediments

How significant is the effect of grain size on chemical composition in bottom sediments? Table 2.2.2-1 contains data for a marine sample that was broken down into various size fractions; each fraction was then subjected to chemical analysis. The sediment displays a bimodal grain-size distribution with peaks in the ranges of <2 μm and 10 to 20 μm; this is typical for marine material (Rex and Goldberg, 1958).

For example, the <2-μm fraction has, by far, the highest Cu concentration of all the fractions. Although the <2-μm fraction represents only some 20 percent of the bulk sediment, its Cu contribution amounts to 75 percent of the total Cu in the sample. The 10 to 20-μm fractional contribution amounts to only 9 percent of the total Cu, although it is the largest single size fraction in the sample. This type of pattern is similar to ones for other trace elements such as Cd, Ni, Co, Zn, and Pb.

2.0 PHYSICAL AND CHEMICAL FACTORS AFFECTING SEDIMENT-TRACE ELEMENT CHEMISTRY

2.2 Physical factors

2.2.3 Chemical analysis of various grain sizes in suspended sediments

Are the patterns just described for bottom sediments also typical of suspended matter? Data presented in Figure 2.2.3-1 should clarify this point. All the trace elements investigated—Mn, Fe, Cr, Ni, Cu, and Co—have their highest concentrations in the <2-μm fraction. Despite this, the figure also shows that the majority of the trace element transport can be attributed to the 2-μm fraction. This occurred because, although the trace element concentrations associated with the size fractions finer than 2 μm were significantly higher than those found in the 2-μm fraction, the mass of sediment in these finer fractions was relatively small, compared to the mass of the 2-μm fraction. Thus, the transport of selected size-fraction sediment-associated trace elements depends on both chemical concentration as well as size-fraction sediment masses. These data could also imply that most *chemical* transport may occur under relatively low discharge/velocity conditions because high trace element concentrations are associated with fine-grained material, which, in turn, dominates suspended sediments during low-flow.

Table 2.2.2-1. Distribution of Copper by Size Fraction in a Bottom Sediment Sample from the Pacific Ocean

Size Fraction (μm)	Percent of Sample	Chemical Concentration (mg/kg)	Contribution to Sample* (mg/kg)
<2	20	750	150
2 - 6	15	60	9
6 - 10	18	110	19.8
10 - 20	30	60	18
20 - 32	10	25	2.5
32 - 64	7	20	1.4
	100		200.7

* - Calculated by multiplying chemical concentration by percent of sample.

Figure 2.2.3-1. Distribution of Trace Elements by Particle Size and Relative Mass Transport by Suspended Sediment in the Amazon and Yukon Rivers (Data from Gibbs, 1977)

EXPLANATION:
Relative mass transport -- calculated from size distribution data of material transported by each river and the trace element concentrations for each size fraction: as a relative term, it has no units.

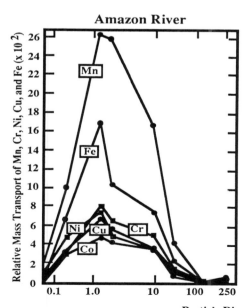

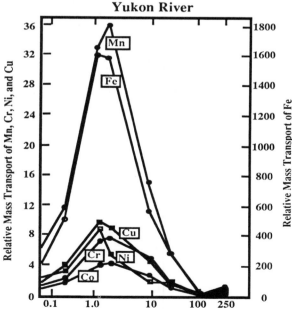

Particle Diameter, in Micrometers

19

2.0 PHYSICAL AND CHEMICAL FACTORS AFFECTING SEDIMENT-TRACE ELEMENT CHEMISTRY

2.2 *Physical factors*

2.2.4 Effect of grain size on trace element concentrations in samples collected from the same and different basins

The effect of grain size on sediment-trace element concentrations should be evident in light of the preceding discussions. Further, it should be apparent that the finer grain sizes can contain the majority of the trace elements associated with sediments. However, it is important to understand the significant implications such interrelations have for such common practices as trying to trace the extent of pollutant discharge from a point source, and for comparing sediment-chemical concentrations within and between depositional basins.

For example, the data in Figure 2.2.4-1 come from bottom sediments collected from the River Ems in Europe (de Groot, et al., 1982). The Mn concentrations cover a wide range, from 700 to about 2500 mg/kg; Co, Fe, and Hg also show wide concentration ranges. Solely on the basis of bulk chemical analysis, there appears to be no interrelation between the various samples, and all that an investigator could report would be the wide range of concentrations. However, when the chemical data are plotted against grain-size information, very distinct patterns emerge. Obviously, there is a strong positive correlation between increasing chemical concentration and the increasing percentage of fine-grained material; however, note that the slopes of the lines for each of the elements are different, which implies that trace element-grain size relations differ for each element. This interrelation (positive correlation between decreasing grain size and increasing trace element concentration), which is common in sediments, can provide a means for tracing the extent of pollutant transport or dispersion from a point source because, as material moves away from the source, it is usually diluted by other constituents and other grain sizes. As a result, the use of bulk chemical data would make it difficult to trace the extent of dispersion because of the effects of dilution. However, these dilution effects could be reduced, and the extent of transport determined, either by separating out the constituent-bearing size fraction or by determining the percentage of the constituent-bearing size fraction in a bulk sediment sample and recalculating the bulk data accordingly. Recalculation methods are discussed in Section 3.0 of this Primer.

The pattern displayed by the River Ems sediments, a strong positive correlation between trace element concentrations and the percentage of <16-µm material, should not be viewed as an endorsement for the universal use of the <16-µm fraction to reduce or eliminate grain-size effects. Studies in other areas have indicated a wide variety of grain-size fractions may be useful in dealing with the grain-size effect. Recommendations cover a broad range of size fractions including: <2µm, <16µm, <20µm, <63µm, <70µm, <125 µm, and <200µm; mean and median grain sizes have been proposed as well (Banat, et al., 1972; Copeland, 1972; Renzoni, et al., 1973; Cameron, 1974; Chester and Stoner, 1975; Webb, 1978; Jenne, et al., 1980; deGroot, et al, 1982; Ackermann, et al., 1983; Beeson, 1984; Horowitz and Elrick, 1987; Horowitz and Elrick, 1988). A further discussion of the selection of the most appropriate grain-size fraction or range can be found in Section 2.4 and 3.3 of this Primer.

Figure 2.2.4-1. Interrelation Between Trace Element Concentration and the Percent <16-μm Fraction for the River Ems, FRG (Data from deGroot, et al., 1982)

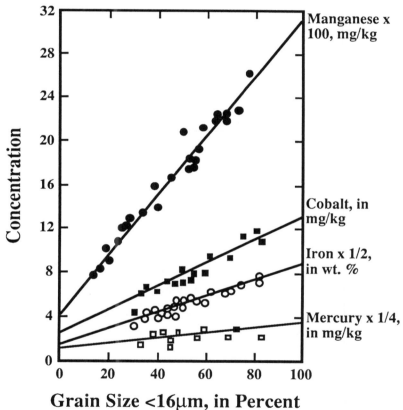

2.0 Physical and Chemical Factors Affecting Sediment-Trace Element Chemistry

2.2 Physical factors

2.2.5 Comparison of samples having similar bulk chemistries but differing grain-size distributions

Bear in mind that two sediments with the same bulk chemical concentrations are not necessarily the same. The sources for the constituents in question may be different, the constituents may be held in different chemical forms, they may be associated with different size fractions, etc. As an example, examine the data in Table 2.2.5-1. Both samples are marine in origin. Sample AD 600 comes from a hydrothermally active area, while sample AD 500 comes from a non-hydrothermal area. Both samples have similar Mn concentrations (~3500 mg/kg) and similar grain-size distributions. However, the chemical data for each grain-size range indicate significant differences between the two samples in the distribution of Mn. Sample AD 600 has two major Mn peaks, one in the <2-µm fraction, and the other in the 10-to-20-µm fraction. Sample AD 500, in contrast, has no Mn peak in any of the size fractions ranging from 2 µm to 32 µm. Subsequent mineralogical and chemical analyses have indicated that manganese oxide coatings (<2 µm) and manganese micro-nodules (10 - 20 µm) caused the Mn peaks in AD 600, while the Mn in AD 500 came from locally derived and aeolian transported volcanic ash and glass. Differentiation of the two samples could not have been made solely on the basis of either grain-size distribution or bulk chemical data, but could be made on the basis of size-fraction chemical analyses.

2.0 Physical and Chemical Factors Affecting Sediment-Trace Element Chemistry

2.2 Physical factors

2.2.6 Effect of grain size on sediment-associated chemical transport at differing discharge rates

It has now been reasonably well established that trace elements tend to be concentrated in the finer grain sizes of both suspended and bottom sediments. It also has been shown that larger-sized material can act as a diluent for the constituent-rich finer grain sizes. What, then, is the implication for suspended-sediment constituent concentrations under differing discharge conditions? The immediate assumption might be that as both discharge and the suspended sediment load increase, associated trace element concentrations also increase. However, a little thought and an examination of the data presented in Figure 2.2.6-1 should dispel this assumption. As discharge increases, the size of sediment grains being suspended and transported will also increase. When fine-grained material, rich in trace elements, is diluted with relatively trace element-poor, larger-sized material, suspended-sediment trace element concentrations decrease, as in the case of Cr (Fig. 2.2.6-1). Remember that the mass of Cr, or other trace elements transported may not decrease and may, in fact, increase (see Table 1.4-1 for calculations concerning transport); only the concentration in the suspended load will decrease (This interrelation is applicable to large rivers; caution should be exercised in applying it to small rivers; further discussion of the relations between discharge, concentration, and transport can be found in Section 4.3.2.4 in this Primer). The data presented in Figure 2.2.6-1, and as also shown in Figure 2.2.3-1, reemphasize the point that most chemical transport may occur under relatively low flow conditions, because of the high trace element concentrations associated with small size fractions that do not require high velocities (discharge) to move them.

Table 2.2.5-1. Comparison of Manganese Distributions in Two Samples Having Similar Grain-Size Distributions and Bulk Concentrations

Size Fraction (μm)	AD-600 Percent of Sample	AD-600 Conc.[1] (mg/kg)	AD-600 Cont.[2] (mg/kg)	AD-500 Percent of Sample	AD-500 Conc.[1] (mg/kg)	AD-500 Cont.[2] (mg/kg)
<2	20	5000	1000	18	4000	720
2 - 6	12	1000	120	13	3800	494
6 - 10	14	1000	140	15	3800	570
10 - 20	30	7100	2130	25	4000	1000
20 - 32	10	1000	100	15	4000	600
32 - 64	7	500	35	5	2000	100
64 - 125	4	300	12	5	1000	50
125 - 250	3	300	9	4	500	20
	100		3550	100		3550

[1]Conc. - concentration in milligrams per kilogram
[2]Cont. - contribution to sample in milligrams per kilogram, calculated by multiplying the chemical concentration by the percent of sample

Figure 2.2.6-1. Effect of Stream Discharge on Chromium Concentration for Suspended Sediments from the Rhine River, FRG (Data from Forstner and Wittmann, 1981)

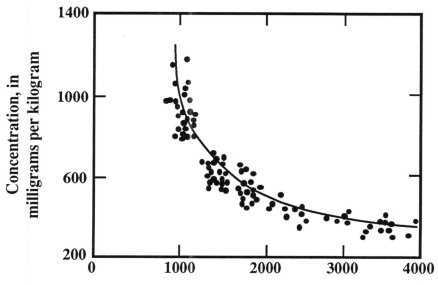

Discharge, in cubic meters per second

2.0 Physical and Chemical Factors Affecting Sediment-Trace Element Chemistry

2.2 Physical factors

2.2.7 Measuring grain-size distributions

By now, it should be obvious that grain size is one of the most significant factors controlling a sediment's capacity for collecting and concentrating trace elements. It should also be apparent that during the course of various water-quality and geochemical studies, it may be necessary to either separate out specific grain-size fractions for subsequent chemical analysis and/or to determine the grain-size distributions of various sediment samples. Numerous techniques are available to carry out either procedure (measurement or separation); unfortunately, comparisons between these various techniques usually display marked differences (e.g., Krumbein and Pettijohn, 1938; U.S. Interagency Committee on Water Resources, 1941; Guy, 1969; Jackson, 1979; Stockham and Fochtman, 1979; Horowitz and Elrick, 1986; Horowitz, 1990). Thus, a grain-size analysis falls into the category of an operationally-defined property. In other words, there are a number of techniques available, there is no single standard technique, and each method can produce a different set of results (Todd, et al., 1963; Horowitz and Elrick, 1986; Horowitz, 1990). This is particularly true when dealing with material finer than 63 μm (the silt/clay fraction).

Material coarser than 63 μm is readily sized by sieving; should subsequent trace element analyses of the separated fractions be required, contamination from metallic sieves (especially from the solder used to attach the sieve mesh to a metallic frame) can be eliminated by using non-metallic ones. Stainless steel sieves, in which the mesh is held in place by crimping, rather than by solder, may also be acceptable.

Grain-size analyses for material finer than 63 μm fall into two distinct categories: those that involve a direct measurement of the grain sizes present and those that involve an actual physical separation (Table 2.2.7-1). The basic assumptions for either type of procedure are engendered in Stokes Law which deals with the fall times of particles in a specific medium (e.g. Krumbein and Pettijohn, 1938).

$$t = \frac{18\mu h}{g(p_p - p_l)d^2}$$

where

t = time, in seconds,
μ = dynamic or absolute viscosity of settling media, in poises,
h = depth of fall, in centimeters,
g = gravitational constant, in centimeters/square second,
p_p = mass density of particle, in grams/cubic centimeter,
p_l = mass density of settling medium, in grams/cubic centimeter, and
d = particle diameter, in centimeters.

Stokes Law assumes that: particles are spheres, they all have the same density, they are smooth and rigid, the settling medium is homogeneous in comparison to the size of the particle, the particle must fall as it would in a medium of unlimited extent, fall occurs at a constant settling velocity, and there are no gravitational effects between particles (in other words, the sediment/settling medium mixture is sufficiently dilute). Few if any sediments can meet all these assumptions; however, Stokes Law techniques are the norm for the grain-size analysis and/or separation of material finer than 63 μm.

Table 2.2.7-1. A Listing of Various Grain Size Analysis Techniques

≥63 μm Fractions
Sieving
Visual Accumulation

≤63 μm Fractions - Physical Separations
Pipet Analysis
Decantation
Bottom Withdrawal
Centrifugation
Air Elutriation

≤63 μm Fractions - Measurements
Electrozone Counters
Sedigraph®
Microscopy
Hydrometers
Image Analyzers

2.0 PHYSICAL AND CHEMICAL FACTORS AFFECTING SEDIMENT-TRACE ELEMENT CHEMISTRY

2.2 *Physical factors*

2.2.7 Measuring grain-size distributions

2.2.7.1 DIFFERENCES IN GRAIN-SIZE DISTRIBUTIONS INDUCED BY USING DIFFERENT TECHNIQUES

As pointed out in the previous section, grain-size analyses can involve a measurement of particle distribution, or they can involve an actual physical separation. Generally, the measurement techniques are more rapid and less time-consuming; data from these types of procedures can be used to normalize sediment-trace element concentrations to deal with the effects of differences in grain-size distributions among a group of samples (data normalization techniques for dealing with the grain-size effect are discussed in Sections 3.3 - 3.6 of this Primer). However, if any subsequent analytical procedures are required for specific size fractions or ranges, then a physical separation must be carried out. This places certain limitations on the separation technique to be selected; thus, the method should meet certain criteria before it can be considered acceptable for subsequent analytical work. First, size separations must provide a sufficiently large sample of each fraction to permit chemical analysis, or any other determination required. Second, it is desirable to carry out the separation in a medium which will not alter the chemical composition of the sample. Third, it is preferable to be able to, for example, chemically analyze individual size fractions of interest directly, rather than to determine their chemical composition by difference, to reduce the cumulative analytical errors associated with multiple chemical analyses.

Remember, grain-size distributions/separations are operationally defined measurements. That means that different procedures may produce markedly different measurements of distribution and/or separated size fractions. The magnitude of the differences can be quite extreme as illustrated in Table 2.2.7.1-1 and Figure 2.2.7.1-1. These differences are of little consequence when comparing a set of samples that have all been treated in the same way; however, they must be borne in mind when comparing data between different studies where the exact same procedures have not been employed.

Table 2.2.7.1-1. Comparison of Size Distributions Obtained by Air Elutriation (Elut.) and Chemically Dispersed Pipet (Pipet) Analyses (Data from Horowitz and Elrick, 1986)

Size Fraction (μm)	Doane Lake Outlet Elut. (%)	Pipet (%)	Swan Island Elut. (%)	Pipet (%)	Lake Bruin Elut. (%)	Pipet (%)	Ned Wilson Lake Elut. (%)	Pipet (%)	Mississippi River Elut. (%)	Pipet (%)
<2	13	29	12	26	12	45	11	47	15	45
2- 4	8	7	11	8	10	5	11	5	12	8
4- 8	11	9	13	10	9	5	15	9	12	11
8-16	19	14	19	16	15	7	25	13	18	16
16-32	12	24	14	21	14	20	17	15	12	16
32-63	36	17	30	20	39	18	22	11	31	4

Figure 2.2.7.1-1. Comparison of Size Fractions Produced by Air Elutriation and Pipet Analysis (Data from Horowitz and Elrick, 1986)

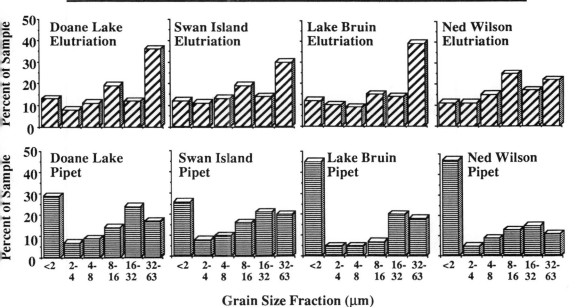

27

2.0 PHYSICAL AND CHEMICAL FACTORS AFFECTING SEDIMENT-TRACE ELEMENT CHEMISTRY

2.2 *Physical factors*

2.2.7 Measuring grain-size distributions

2.2.7.2 DIFFERENCES IN GRAIN-SIZE DISTRIBUTIONS INDUCED BY SAMPLE PRETREATMENT

As demonstrated in the previous section, different grain-size analytical techniques can produce markedly different measures of distribution and/or separated size fractions. Many size-separation methods require some form of sample pretreatment prior to the actual separation. Sample pretreatments can be of particular concern with respect to any subsequent chemical analyses of the separated material. Many aqueous size-separation techniques require the use of some type of dispersing agent such that a homogeneous mixture is obtained (e.g., Guy, 1969). One of the most commonly employed dispersing agents is sodium hexametaphosphate (Calgon®). Unfortunately, this dispersing agent can alter the chemical composition of the separated material by adding Na to and/or by displacing various trace elements from the separated size fractions. Further, many techniques also require a pretreatment with hydrogen peroxide to destroy organic matter (e.g., Guy, 1969). This also may be an unacceptable procedure because many trace elements are associated with the organic fraction of a sediment sample (see Sections 2.3.2 and 2.3.2.2 in this Primer).

Many size-separation techniques which usually call for some type of sample pretreatment can be performed without the use of the pretreatment step(s). However, it must be clearly understood that the elimination of any pretreatment step(s) may produce an entirely different size analysis than the same procedure which included the pretreatment. In other words, the elimination of the pretreatment may cause a substantial change in the operational definition of the sizing technique. The effects of omitting pretreatment are amply illustrated in Table 2.2.7.2-1 and Figure 2.2.7.2-1. As with the variations that can occur when alternate size analysis procedures are employed, these differences are of little consequence when comparing a set of samples that have all been treated in the same way; however, they must be borne in mind when comparing data between different studies where the same procedures have not been employed.

Table 2.2.7.2-1. Comparison of Size Distributions Obtained by Chemically Dispersed (P/CD) and Non-chemically Dispersed (Pipet) Analyses (Data from Horowitz and Elrick, 1986)

Size Fraction (μm)	Doane Lake Outlet P/CD (%)	Doane Lake Outlet Pipet (%)	Swan Island P/CD (%)	Swan Island Pipet (%)	Lake Bruin P/CD (%)	Lake Bruin Pipet (%)	Ned Wilson Lake P/CD (%)	Ned Wilson Lake Pipet (%)	Mississippi River P/CD (%)	Mississippi River Pipet (%)
<2	29	13	26	10	45	11	47	5	45	21
2- 4	7	7	8	10	5	9	5	7	8	12
4- 8	9	14	10	11	5	9	9	11	11	14
8-16	14	21	16	21	7	16	13	19	16	21
16-32	24	18	21	29	20	28	15	28	16	21
32-63	17	26	20	20	18	27	11	30	4	11

Figure 2.2.7.2-1. Comparison of Size Fractions Produced by Pipet Analysis Using Chemical Dispersion (Chem. Disp.) and Native Water (Nat. Wat.) (Data from Love, 1960)

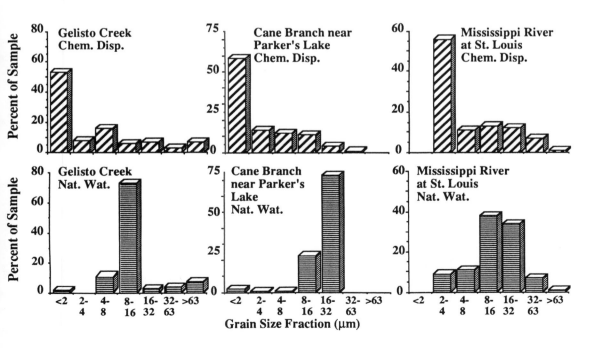

2.0 PHYSICAL AND CHEMICAL FACTORS AFFECTING SEDIMENT-TRACE ELEMENT CHEMISTRY

2.2 Physical factors

2.2.8 Effect of sediment surface area

Surface area is almost indistinguishable from grain size; as grain size decreases, surface area increases sharply. This would not be true if surface area was dependent solely on an individual grain of a given diameter. Consider the surface area of two spheres, one having a diameter (grain size) of 100 μm and the other having a diameter of 10 μm. Based on the mathematical formula for the surface of a sphere (πD^2), the 100-μm particle would have a surface area of 31,416 μm^2 while the 10-μm particle would have a surface area of only 314 μm^2. However, surface area is reported as area per mass. A given mass of sediment will contain many times more finer-diameter particles than coarser-diameter ones (Fig. 2.2. 8-1). Hence the surface area of sediments increases with decreasing grain size.

As an example, examine the data in Table 2.2.8-1. The surface areas for simple spheres of differing diameters (grain size) have been calculated (Jackson, 1979). Sand-sized particles have surface areas on the order of tens to hundreds of cm^2/g, silt-sized particles have surface areas on the order of hundreds of thousands of cm^2/g, and clay-sized particles have surface areas on the order of tens of m^2/g. Bear in mind that these numbers are for *spheres*. Obviously, naturally occurring sediments are composed not of spheres, but of irregular grains. Jackson (1979) has calculated that the surface area of montmorillonite (a platey clay mineral) can theoretically be more than 800 m^2/g (if it were a sphere of 2 μm in diameter, its surface area would only be 1.13 m^2/g).

Actual measurement of naturally occurring materials, including montmorillonite, indicate that surface area can be and usually is less than the theoretical value, but is still much larger than that of a sphere of equivalent diameter (Forstner and Wittmann, 1981). This is amply demonstrated in Table 2.2.8-2 where data for selected materials with diameters all less than 2 μm is presented. These data also illustrate that surface area is a function of both the physical size of a sediment (grain size) as well as composition (e.g., kaolinite versus illite versus montmorillonite which are all clay minerals yet each has markedly different surface areas).

Figure 2.2.8-1. Figure Demonstrating Why Surface Area Increases with Decreasing Grain Size for a Given Mass of Sediment

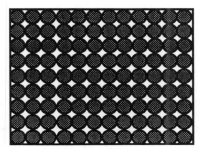   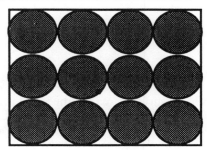

Fine Grain Size   Coarse Grain Size

Table 2.2.8-1. Calculated Surface Areas Assuming Sphericity (Data from Jackson, 1979)

		Surface Area	
Size Class	Diameter (μm)	(m²/g)	(cm²/g)
Very Coarse Sand	2000	0.00113	11.3
Very Fine Sand/Coarse Silt	62	0.036	360
Very Fine Silt/Coarse Clay	4	0.57	5,700
Medium Clay	2	1.13	11,300
Fine Clay	1	2.26	22,640
Very Fine Clay	0.5	4.52	45,280
Colloids	0.1	22.6	226,400

Table 2.2.8-2. Surface Areas of Selected Materials with Diameters <2 μm (Data from Forstner and Wittmann, 1981)

Material	Surface Area (m²/g)
Calcite	12.5
Clay Minerals	
Kaolinite	10 - 50
Illite	30 - 80
Montmorillonite	50 - 150
Iron Hydroxide	300
Organic Matter	1900

2.0 PHYSICAL AND CHEMICAL FACTORS AFFECTING SEDIMENT-TRACE ELEMENT CHEMISTRY

2.2 *Physical factors*

2.2.9 Importance of surface area to sediment-trace element concentrations

Why is surface area so important in controlling sediment-trace element concentrations? It is because all the various means by which sediment tends to collect, concentrate, and retain trace elements fall into the general category of so-called surface reactions or surface chemistry. It follows that materials with large surface areas (small grain sizes) are the main sites for the transport and collection of these constituents (Krauskopf, 1956; Jenne, 1968; Gibbs, 1973; Jenne, 1976; Jones and Bowser, 1978; Jenne, et al., 1980; Forstner and Wittmann, 1981; Horowitz and Elrick, 1987). The major mechanism for the collection of trace elements on surfaces is adsorption. This process entails the condensation of atoms, ions, or molecules on the *surface* of another substance. Materials having large surface areas are good adsorbers. Adsorption can occur with or without cation exchange (see Section 2.3.1), and should not be confused with absorption, which involves penetration of a substance into the body, or inner structure, of another material. Additionally, Jenne (1976) indicates that materials with large surface areas can be viewed simply as mechanical substrates upon which inorganic constituents can concentrate without any chemical interaction between the material and the constituent. Thus, deposited materials like organic matter, and hydrous iron and manganese oxides, rather than the original surface, may act as a trace element collector. Obviously, as surface area increases, so does the amount of these collectors, thus increasing the trace element-concentrating capacity of the surface.

An example of how surface area affects trace element concentrations is shown in Figure 2.2.9-1. The data come from the Ottawa and Rideau Rivers and clearly show that as surface area increases, the concentrations of both Ni and Cu also increase (Oliver, 1973). Please note that the relation between surface area and trace element concentration is not linear, but is probably logarithmic (e.g., Horowitz and Elrick, 1987). This pattern of increasing trace element concentration with increasing surface area also is typical for many other trace elements such as Fe, Mn, Zn, Pb, As, Cr, and Hg (Horowitz and Elrick, 1987). Also, note that the surface areas plotted in these figures (Fig. 2.2.9-1) are actual measurements—they are not calculated. Therefore, do not assume, for example, that a surface area of 22.6 m^2/g implies a sediment particle size of some 0.1 μm, which would be the case if the particle were a sphere (see Section 2.2.8).

**Figure 2.2.9-1. Effect of Sediment Surface
Area on Trace Element Concentrations
from the Ottawa and Rideau Rivers,
Canada (Data from Oliver, 1973)**

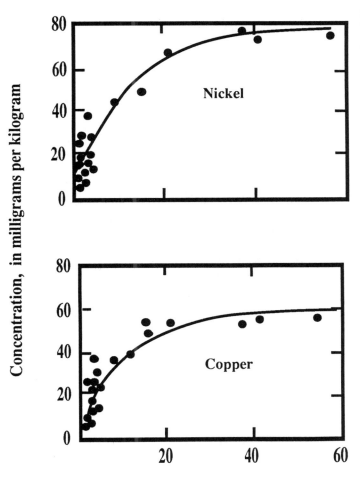

2.0 PHYSICAL AND CHEMICAL FACTORS AFFECTING SEDIMENT-TRACE ELEMENT CHEMISTRY

2.2 Physical factors

2.2.10 Measuring surface area

Based on the foregoing discussions, surface area obviously is a significant factor in controlling the capacity of a sediment to collect, concentrate, and retain trace elements. The accurate determination of the surface area of minerals, soils, and sediments commenced in 1938 with the use of gas adsorption techniques and the development of the BET (Brunauer, Emmett, Teller) equations (Brunauer, et al., 1938). Today, the same procedures are still used, but there have been substantial improvements in the instrumentation used to make the measurements.

As with grain-size measurements, surface area also is an operationally defined property; there are numerous techniques available for its determination (Table 2.2.10-1). As such, surface area determinations using different techniques will produce different values for the same material (Table 2.2.10-2). In addition, there are substantial variations for each technique depending on the type of gas used, or whether a single-point or multi-point procedure is employed to measure the sorptive capacity of a sample. Mixtures of nitrogen and helium gas are the most commonly employed adsorbant, with adsorption taking place at a temperature of -196°C (in a liquid nitrogen bath), and desorption taking place at room temperature (e.g., Micromeritics, 1986). However, numerous other gases such as argon, krypton, carbon dioxide, and n-butane have been mixed with helium for measuring surface area (e.g., Micromeritics, 1986). The different gas mixtures require adsorption at different temperatures than the nitrogen/helium mixture. In addition, other techniques employing various liquid adsorbants and dyes also have been used (Table 2.2.10-1; Jackson, 1979). In general, the gas techniques measure what has been termed the 'external' surface area of a solid because the gas molecules are usually too large to enter the interstices of, for example, expandable clay minerals such as montmorillonite. On the other hand, most liquids and dyes do enter mineral interstices and thus, provide a measure of 'total' (external plus internal) surface area.

Surface area has been shown to correlate positively with decreasing grain size (Horowitz and Elrick, 1987). As such, it could be used in lieu of grain-size determinations to normalize sediment-trace element data to eliminate the 'grain-size effect'. One of the major reasons for using surface-area measurements instead of grain-size data results because it has been demonstrated that gas adsorption surface area measurement techniques are both non-contaminating and non-destructive (Horowitz and Elrick, 1987). This may not be important in some situations; however, when sample sizes are relatively small, and a number of different analyses are required, the need to sample mass while obtaining useful information, may dictate the use of surface-area data instead of grain-size data.

Table 2.2.10-1. Methods for Determining Sediment Surface Area

Gas Adsorption Techniques - Using Various Gas Mixtures

Single-Point Static Gas Adsorption   Nitrogen/Helium

Multipoint Static Gas Adsoprtion   Argon/Helium

Single-Point Flowing Gas Adsorption   Krypton/Helium

Multipoint Flowing Gas Adsorption   n-Butane/Helium

Liquid/Dye Adsorption Techniques - Using Various Solutions

Water

Ethylene Glycol

Glycerol

Ammonium Chloride

Table 2.2.10-2. Comparison of Sediment Surface Area Measurement Techniques (Data from Horowitz, 1990)[1]

Single-Point Flowing Gas Technique

NIST RM 8570 : calcined kaolin

Certified by Round Robin Testing Using ASTM Method 3663

Single-Point Flowing Gas Certified Surface Area = 10.3 m^2/g

Single-Point Flowing Gas Surface Area = 10.5 $\pm$ 0.2 m^2/g (n = 10)

Static Gas vs. Single-Point Flowing Gas Techniques

NIST CRM M11-08 : α- alumina

Static Gas Technique Certified Surface Area = 0.780 $\pm$ 0.004 m^2/g

Single-Point Flowing Gas Surface Area = 0.68 $\pm$ 0.01 m^2/g (n = 10)

[1]Where the methods are the same, as for the calcined kaolin, the mean for 10 separate determinations is within analytical error, where the methods differ, as with the alumina, the mean differs from the certified value by significantly more than the analytical error.

2.0 PHYSICAL AND CHEMICAL FACTORS AFFECTING SEDIMENT-TRACE ELEMENT CHEMISTRY

2.3 Chemical factors

Now that we have discussed the major physical factors affecting the interactions between trace elements and sediments, as well as physical partitioning, it is time to examine the chemical factors affecting the interactions between trace elements and sediments, as well as chemical partitioning. Just as aquatic systems are physically dynamic (water moving in a streambed, wind-created waves in a lake) they also are chemically dynamic. The key to understanding and predicting trace element transport and environmental availability, as well as to identifying sources and sinks for aquatic trace element constituents, is the identification and quantitation of the trace element associations on/in sediments (suspended and bottom) and the reactions among sediment, water, and biota.

A search of the sediment-chemical literature dealing with the chemical factors controlling sediment-trace element associations indicates that two approaches have been used. The first approach aims at determining how trace elements are retained on/in sediments—the so-called mechanistic approach. According to Gibbs (1977) there are five major mechanisms for trace element accumulation on/in sediments (Table 2.3-1). Adsorption was discussed in Section 2.2.9. Precipitation and coprecipitation are readily understandable terms. In aquatic systems both mechanisms tend to occur, *in situ*, and produce authigenic solid phases. Organometallic bonding is the attachment of a trace element directly to carbon; such bonds can be formed by almost any trace element and by several major elements (e.g., phosphorus, silicon). Examples are diethylzinc, methyl- and dimethyl mercury, and tetraethyl lead. Some organometallics can occur naturally, *in situ*, either chemically or with some biological mediation (e.g., bacterial action producing methyl- and dimethyl mercury) while others may have an anthropogenic origin (e.g., tetraethyl lead from leaded gasoline). Incorporation in crystalline minerals is called substitution. This entails the replacement of one element for another within a fixed crystal structure and is more common in solid solutions than in water-solid interactions. Substitution is governed by ionic radius and charge (e.g., Goldschmidt's Rules, Ramberg's Rules, both of which are empirical). In general— and there are exceptions—substitution may take place when ionic radii are within ±15 percent and when the charge is the same, or differs by no more than ±1. For example, of the three cations: Na (charge = +1, ionic radius = 0.98Å), K (charge = +1, ionic radius = 1.33Å), and Ca (charge = +2, ionic radius = 1.06Å), Na could substitute for Ca but not K, and K could not substitute for either Na or Ca.

The second approach seeks to determine where trace elements are retained on/in sediments (phase or site)—the so-called phase approach (Table 2.3-2). This approach has been attempted because individual constituents (e.g., Fe, Mn, Cd, Zn, Ni, Co, Cd) may be, and usually are, associated with several phases. The term *phase* is used in the thermodynamic sense and incorporates categories like interstitial water, clay minerals, sulfides, carbonates, organic matter, iron oxides, etc. (Table 2.3-2). Despite this relatively simple division into two approaches, very few attempts to chemically partition complex sediment samples entail a purely mechanistic or phase approach; rather, they combine aspects of both.

**Table 2.3-1. Mechanisms for Trace Element
Accumulation in Sediments (Data from Gibbs, 1977)**

- Adsorption on Fine-Grained Material
- Precipitation of Trace Element Compounds
- Coprecipitation with Hydrous Iron and Manganese
  Oxides and Carbonates
- Association Either by Adsorption or Organometallic
  Bonding with Organic Matter
- Incorporation in Crystalline Minerals

**Table 2.3-2. Examples of Trace Element
Accumulative Phases in Sediments**

- Interstitial Water
- Carbonates
- Clay Minerals
- Organic Matter
- Hydrous Iron Oxides
- Manganese Oxides
- Sulfides
- Silicates

2.0 PHYSICAL AND CHEMICAL FACTORS AFFECTING SEDIMENT-TRACE ELEMENT CHEMISTRY

2.3 Chemical factors

2.3.1 Cation exchange capacity

Many different materials that are common components of sediments, and that have large surface areas, such as clay minerals, iron hydroxides, manganese oxides, and organic matter, are capable of sorbing cations from solution and releasing equivalent amounts of other cations back into solution. This process is called ion exchange. Because most trace elements behave as cations (have a positive charge), and the surfaces of most materials with a capacity for this process have a net negative charge, the process is termed cation exchange. The capacity of a particular material to hold cations can be measured and is called CEC (cation exchange capacity). The actual mechanism by which cation exchange occurs is a matter of much research and debate, but appears to be due to the sorptive properties of negatively charged anionic sites such as $SiOH$, $Al(OH)_2$, and $AlOH$ (clay minerals), $FeOH$ (iron hydroxides), and $COOH$ and OH (organic matter) (Forstner and Wittmann, 1981). CEC may also occur between the layers of certain silicates, such as expandable clays (montmorillonite, smectite, vermiculite), which, depending on how one chooses to define the layers, might not be viewed as 'surfaces' (Grim, 1968).

CEC, like many other sedimentary properties, is operationally defined. It is usually determined by measuring the concentration of a particular cation that a sample can remove from a solution once the material and the solution have reached equilibrium (e.g., Jackson, 1979). If the solute in the solution is changed, then the CEC also changes. Typical solutes include Ca, K, Na, and NH_4 (Jackson, 1979). CEC is normally reported as milliequivalents of solute per hundred grams of solid (mEq/100 g).

The data in Table 2.3.1-1 include measured CECs for selected materials commonly associated with this phenomenon. The data also show that as grain size decreases and surface area increases, CEC increases substantially (Tables 2.3.1-2 and 2.3.1-3). Note that CEC results from the availability of 'unbalanced' negatively charged sites; thus, it could readily be stated that it is the result of a net negative surface charge [an important physical factor (see Section 2.1)]. The amount and sign (positive or negative) of charge on a solid surface is called its zeta potential.

Table 2.3.1-1. Cation Exchange Capacity of Selected Materials

Material	Exchange Capacity mEq/100g	Reference
Kaolinite	3 - 15	1
Illite	10 - 40	1, 2
Chlorite	20 - 50	1
Montmorillonite	80 - 120	1, 2
Smectites	80 - 150	2
Vermiculites	120 - 200	2
Iron Hydroxide	10 - 25	1
Soil Humic Acids	170 - 590	1
Manganese Oxides	200 - 300	1

1 - From Forstner and Wittmann (1981)
2 - From Drever (1982)

Table 2.3.1-2. Variation of Cation Exchange Capacity with Particle Size for Kaolinite (Data from Grim, 1968)

Particle Size (μm)	10 - 20	5 - 10	2 - 4	1 - 0.5	0.5 - 0.25	0.25 - 0.1	0.1 - 0.05
Cation Exchange Capacity (mEq/100g)	2.4	2.6	3.6	3.8	3.9	5.4	9.5

Table 2.3.1-3. Variation in Cation Exchange Capacity with Particle Size for Illite (Data from Grim, 1968)

Particle Size (μm)		1 - 0.1	0.1 - 0.06	<0.06
Cation Exchange Capacity (mEq/100g)	Sample A	18.5	21.6	33
	Sample B	13.0	20.0	27.5
	Sample C	20.0	30.0	41.7

2.0 PHYSICAL AND CHEMICAL FACTORS AFFECTING SEDIMENT-TRACE ELEMENT CHEMISTRY

2.3 Chemical factors

2.3.2 Composition—significant sedimentary trace element collectors

As has already been shown (see Section 2.2), a number of physical factors affect the capacity of a sediment to attract and concentrate trace elements. With these factors as a basis, plus additional chemical work, it is possible to identify the compounds and substances (geochemical substrates) that are most important in providing a sediment with the capacity to concentrate trace elements (Table 2.3.2-1). These geochemical substrates have large surface areas, high cation exchange capacities, and high surface charges, and they all tend to be concentrated in the finer size fractions. Further, sedimentary trace element collectors tend to be thermodynamically unstable and are amorphous or cryptocrystalline (Jones and Bowser, 1978). The most common materials meeting these criteria are hydrous manganese oxides, and hydrous iron oxides, organic matter, and clay minerals (Table 2.3.2-1).

Forstner (1982a) has listed various geochemical substrates in descending order according to their capacity to collect and concentrate trace elements (Table 2.3.2-1). These results are based upon sequential extraction studies (see Section 2.3.4.2). Within each category there is substantial variability depending upon the geochemical character of the environment (type of clay minerals present, concentration of organic matter, pH, Eh) and the various trace elements involved. Horowitz and Elrick (1987) have also listed various geochemical substrates in descending order according to their capacity to collect and concentrate trace elements (Table 2.3.2-1). These results are based on a hierarchical ranking of correlation coefficients between the geochemical substrates and various trace elements. This listing is obviously different from the one provided by Forstner (1982a). The differences are due to a combination of factors including the source of the sediments (Forstner's data come from marine material while Horowitz and Elrick's data come from freshwater material) and the procedures used to define the various geochemical substrates (partitioning methods employed different operational definitions, see Section 2.3.4.2). Neither set of rankings is correct or incorrect, but they are different. These differences between the rankings of the various geochemical substrates emphasizes some of the difficulties inherent in comparing operationally defined data, as well as comparing data that come from different environments.

Table 2.3.2-1. Compositional Controls on Trace Element Concentration

Important Features

Thermodynamically Unstable

Amorphous or Cryptocrystalline

Capable of Extensive Substitution

High Cation Exchange Capacity

Large Surface Area

High Surface Charge

Small Particle Size

Most Significant Collectors

Hydrous Iron Oxides

Hydrous Manganese Oxides

Organic Matter

Clay Minerals

Relative Capacity of Collectors (Data from Forstner, 1982a)

Manganese Oxides > Organic Matter >

Iron Oxides > Clay Minerals

Relative Capacity of Collectors (Data from Horowitz and Elrick, 1987)

Amorphous Iron Oxides > Total Extractable Iron

> Total Organic Carbon > Reactive Iron > Clay

Minerals > Total Extractable Manganese >
Manganese Oxides

2.0 PHYSICAL AND CHEMICAL FACTORS AFFECTING SEDIMENT-TRACE ELEMENT CHEMISTRY

2.3 Chemical factors

2.3.2 Composition—significant sedimentary trace element collectors

2.3.2.1 IRON AND MANGANESE OXIDES

The significant characteristics of iron and manganese oxides are outlined in Table 2.3.2.1-1. These substances have long been known as excellent scavengers of trace elements from solution (Goldberg, 1954; Krauskopf, 1956). Although the most spectacular demonstration of their significance is the manganese nodules located at the sediment-water interface on deep ocean floors or on lake beds (Mero, 1962; Moore, et al., 1973), the separation and identification of manganese micronodules and iron and manganese oxide coatings on mineral grains in core and grab samples indicate that they are ubiquitous and play an important role throughout the sediment column as trace element collectors in aquatic environments (Goldberg and Arrhenius, 1958; Chester and Hughes, 1967; Jenne, 1968; Cronan and Garrett, 1973; Duchart, et al., 1973; Dymond, et al., 1973; Moore, et al., 1973; Horowitz, 1974; Lee, 1975; Horowitz and Cronan, 1976; Jones and Bowser, 1978; Forstner, 1982a, b; Horowitz and Elrick, 1987). Micronodules have been found in a variety of size ranges but, as their name implies, tend to be concentrated in the smaller (<20 µm) size ranges (Goldberg and Arrhenius, 1958; Chester and Hughes, 1969; Dymond, et al., 1973; Horowitz, 1974).

In soils, suspended sediments, and in bottom sediments, iron and manganese oxides also commonly occur as coatings on various minerals and finely dispersed particles (Forstner and Wittmann, 1981; Salomons and Forstner, 1984; Horowitz and Elrick, 1987). Those forms most capable of concentrating trace elements range from amorphous to microcrystalline to crystalline and have large surface areas—on the order of 200 to 300 m^2/g (Fripiat and Gastuche, 1952; Buser and Graf, 1955). Regardless of form, whether micronodules or coatings, hydrous iron and manganese oxides are important concentrators for trace elements in aquatic systems.

Table 2.3.2.1-1. Physical and Chemical Characteristics of Iron and Manganese Oxides

- Fine Grained

- Amorphous or Poorly Crystallized

- Large Surface Area

- High Cation Exchange Capacity

- High Negative Surface Charge

2.0 PHYSICAL AND CHEMICAL FACTORS AFFECTING SEDIMENT-TRACE ELEMENT CHEMISTRY

2.3 Chemical factors

2.3.2 Composition—significant sedimentary trace element collectors

2.3.2.2 ORGANIC MATTER

The characteristics of aquatic and soil organic matter are outlined in Table 2.3.2.2-1. The capacity of organic matter to concentrate trace elements in and on soils as well as suspended and bottom sediments is well recognized (Goldberg, 1954; Krauskopf, 1956; Kononova, 1966; Swanson, et al., 1966; Saxby, 1969; Schnitzer and Kahn, 1972; Gibbs, 1973; Bunzl, et al., 1976; Jenne, 1976; Singer, 1977; Stoffers, et al., 1977; Nriagu and Coker, 1980; Ghosh and Schnitzer, 1981; Forstner, 1982a, b; Horowitz and Elrick, 1987; Hirner, et al., 1990). Aquatic organic matter, generally termed humic substances, has been subdivided by Jonasson (1977) into four categories: humins, humic acids, fulvic acids, and yellow organic acids (Table 2.3.2.2-1).

Gibbs (1973), among others, has indicated the importance of organic molecules in controlling trace element concentrations on/in suspended and bottom sediments, and in sediment-water interactions. Saxby (1969) has shown that the relative attraction between trace elements with colloidal, suspended, and bottom sediment-associated organic matter can range from weak and readily replaceable (adsorption) to strong (chemically bonded). The capacity of organic matter to concentrate trace elements varies with the constituent and the type of organic matter (Swanson, et al., 1966; Saxby, 1969; Rashid, 1974; Bunzl, et al., 1976; Jonasson, 1977). Organic matter can concentrate between 1 percent and 10 percent dry weight of Co, Cu, Fe, Pb, Mn, Mo, Ni, Ag, V, and Zn (Swanson, et al, 1966; Hirner, et al., 1990). This capacity to concentrate various trace elements appears to be related to several factors, including: large surface area, high cation exchange capacity, high negative surface charge, and physical trapping (Table 2.3.2.2-1). It is also related to the stability of the organo-trace element constituent complex. In soils, the sequence in descending order is represented by the so-called Irving-Williams series (Table 2.3.2.2-1; Irving and Williams, 1948). Similar results have also been found for aquatic organic matter (Swanson, et al., 1966; Saxby, 1969; Rashid, 1974; Bunzl, et al., 1976; Jonasson, 1977).

The concentration of aquatic organic matter, as indicated by such measurements as total organic carbon and total organic nitrogen, tends to show a strong positive correlation with decreasing grain size and increasing surface area (Kuenen, 1965; Forstner and Wittmann, 1981; Salomons and Forstner, 1984; Horowitz and Elrick, 1987). This relatively simple picture is complicated by the fact that aquatic organic matter exists in two physical forms: surface coatings which tend to concentrate in the finer size fractions and separate particles which tend to be associated with the coarser size fractions (Horowitz and Elrick, 1987). Current techniques do not provide very adequate means for differentiating between coatings and particulate organic matter, although some methods are available (e.g., some type of flotation).

Table 2.3.2.2-1. Physical and Chemical Characteristics of Aquatic Organic Matter

Characteristics

- Concentration Increases with Decreasing Grain Size
- Large Surface Area
- High Cation Exchange Capacity
- High Negative Surface Charge
- Capable of Physical Trapping

Types[1]

- Humins
- Humic Acids
- Fulvic Acids
- Yellow Organic Acids

Trace Element Affinities[2]

Lead > Copper > Nickel > Cobalt > Zinc >

Cadmium > Iron > Manganese > Magnesium

[1] - Data from Jonasson (1977)
[2] - Data from Irving and Williams (1948)

45

2.0 PHYSICAL AND CHEMICAL FACTORS AFFECTING SEDIMENT-TRACE ELEMENT CHEMISTRY

2.3 Chemical factors

2.3.2 Composition—significant sedimentary trace element collectors

2.3.2.3 CLAY MINERALS

The physical and chemical characteristics of clay minerals are outlined in Table 2.3.2.3-1. These materials can act as important collectors and concentrators of trace elements in aquatic systems (Goldberg, 1954; Krauskopf, 1956; Goldberg and Arrhenius, 1958; Hirst, 1962; Grim, 1968; Kharkar, et al., 1968; Gibbs, 1973; Jenne, 1976; Jones and Bowser, 1978; Forstner and Wittmann, 1981; Forstner, 1982a, b; Salomons and Forstner, 1984; Horowitz and Elrick, 1987). Clay mineral capacity for cation exchange is governed by broken chemical bonds around the edges of mineral grains, the substitution of Al^{+3} for Si^{+4} with the associated negative charge imbalance, and the presence of expandable lattices (Grim, 1968).

Hirst (1962) has pointed out the importance of different clay minerals in controlling the background levels of trace elements in bottom sediments and has evaluated the capacity of various types of clay minerals with respect to their ability as trace element concentrators (Table 2.3.2.3-1). The actual process by which clay minerals concentrate constituents is not thoroughly understood; however, laboratory studies indicate that it can be rapid (on the order of tens of minutes), and depends upon a number of physicochemical factors such as the valence of the dissolved constituents, their ionic radii, the concentration of the trace element, the type of clay mineral, the pH of the solution, and the nature and concentration of competing substrates (Forstner and Wittmann, 1981; Salomons and Forstner, 1984; Calmano, et al., 1988; Forstner, et al., 1989). An empirically derived affinity sequence for a limited number of trace elements with clay minerals, in descending order, is Pb, Ni, Cu, and Zn (Mitchell, 1964). In other words, if a solution contained all four trace elements, Pb would be preferentially concentrated over Ni, Cu, and Zn by clay minerals; Ni would be preferentially concentrated over Cu and Zn by clay minerals, etc.

The results from some recent studies in some European rivers suggest that clay minerals may not play a significant role in the direct chemical concentration of trace elements by, for example, adsorption (Forstner and Wittmann, 1981). This conclusion has received some support from additional studies carried out in the U.S. (Horowitz and Elrick, 1987). All these results support the view of Jenne (1976) that the major role of clay minerals as trace element concentrators is to act as mechanical substrates for the precipitation and flocculation of organic matter and secondary minerals (e.g., hydrous iron and manganese oxides). In other words, the clay minerals are coated with material(s) that, rather than the minerals themselves, actually carry out the concentration of trace elements.

Table 2.3.2.3-1. Physical and Chemical Characteristics of Clay Minerals

Characteristics

• Fine Grained

• Large Surface Area

• Moderate to High Cation Exchange Capacity

• High Negative Surface Charge

      a. Broken Bonds on Mineral Edges

      b. Substitution of Al^{+3} for Si^{+4}

Capacity for Trace Element Concentration[1]

Montmorillonite > Vermiculite > Illite =

Chlorite > Kaolinite

[1] - Data from Hirst (1962)

2.0 PHYSICAL AND CHEMICAL FACTORS AFFECTING SEDIMENT-TRACE ELEMENT CHEMISTRY

2.3 Chemical factors

2.3.3 Introduction to, and utility of chemical partitioning

The subject of trace element-sediment partitioning entails all the various procedures that can be used to determine how (mechanistic approach) and/or where (phase approach) trace elements are associated with sediments (see Section 2.3). The goal of many water-quality projects, at least initially, is to describe existing conditions at some point in time and to attempt to infer their effect on the environment. However, because aquatic systems are chemically dynamic, the goal of at least some water-quality projects will also entail an attempt to predict likely occurrences later, or the effects of sediment movement downstream from the original study site, or the effect(s) of physicochemical changes on sediment-trace element chemistry. To meet these ends, an investigator must determine how and/or where trace elements are associated with sediments (determine sediment-trace element partitioning).

Table 2.3.3-1 points out some of the potential uses of chemical partitioning and indicates some of the potential changes that can result from altered physicochemical conditions. The table is divided into three major categories: bioavailability, transport modelling, and remobilization. Bioavailability is a common term encompassing a concept that most individuals can grasp. However, determining actual bioavailability is another matter entirely. Bioavailability is always operationally defined, usually as the result of some chemical or toxicological test. Unfortunately, no one has yet developed a universally-accepted method of determining bioavailability, and thus, the concept remains subject to much interpretation and argument (Baker, 1980; Forstner and Wittmann, 1981; Lichtenberg, et al., 1988; Ongley, et al., 1988; Ahlf, et al. 1989; Forstner, 1989).

Because aquatic systems are both physically and chemically dynamic, both aspects must be considered when addressing sediment-associated chemical transport. The physical aspects of sediment transport are, by far, simpler to deal with than the chemical aspects. Physical transport involves hydrodynamics. There are a number of existing sediment transport models and several have been tested and successfully employed to predict sediment movement. Unfortunately, many of these models are applicable only to sand-sized (>63 μm) material and as already discussed, sediment-associated trace element transport is likely to be greatest for the silt/clay fraction (material <63 μm). Further, to predict chemical transport adequately, an effective model must be able to take into account chemical changes that occur during sediment transport due to changing environmental conditions. Today, no such model exists. The first step toward creating such a model is an understanding of all the potential chemical reactions that can take place in an environmental system; further, such a model would also require input regarding existing chemical phases and the way in which trace elements are entrained. Data on existing phases and associations can only be obtained by some form of chemical partitioning.

Under the heading "Remobilization", Table 2.3.3-1 provides practical examples of how changing physicochemical conditions might affect sediment-trace element concentrations (Forstner and Wittmann, 1981; Salomons and Forstner, 1984; Calmano, et al., 1988; Forstner, 1989). The types of changes listed are fairly typical of many common environments and should be familiar occurrences to most water-quality investigators.

Table 2.3.3-1. Various Rationales for the Determination of Chemical Partitioning

I. BIOAVAILABILITY
II. TRANSPORT MODELLING
III. REMOBILIZATION

    A. *Elevated Salt Levels - Alkali and Alkaline Earth Displacement*
        1. Deicing Salts
        2. Saline Effluents
        3. Saltwater Intrusion
        4. Estuarine, Fresh, and Saltwater Mixing
    B. *Redox Changes - Dissolution of Fe and Mn Oxides and Hydroxides with Subsequent Trace Element Release*
        1. Eutrophication
        2. Organic Inputs
    C. *pH Reduction - Dissolution of Carbonates and Hydroxides, Desorption by H^+*
        1. Acid Mine Drainage
        2. Acid Rain
        3. Acidic Industrial Effluents
    D. *Increase in Complexing Agents - Formation of Stable Trace Element Complexes*
        1. Tannic Acid
        2. Peat and Lignite
        3. Nitroacetic Acid (NTA) - Phosphate Replacement
    E. *Microbial Activity - Solubilization of Trace Elements by Direct and Indirect Action*
        1. Alkylation of Trace Elements (e.g., Hg, As, Pb, Se)
        2. Destruction of Organic Matter
        3. Metabolic Activity Decreasing pH or Oxygen

2.0 PHYSICAL AND CHEMICAL FACTORS AFFECTING SEDIMENT-TRACE ELEMENT CHEMISTRY

2.3 Chemical factors

2.3.4 Chemical partitioning methods

Table 2.3.4-1 outlines the most common methods by which chemical partitioning is ascertained. Conceptually, phase chemical partitioning is extremely simple. Imagine taking a sample, placing it under an optical microscope, sifting through the individual sediment grains, identifying the mineralogy of the individual grains, and picking out and placing grains of similar mineralogy in individual piles. Once all this sorting is accomplished, each individual pile can be subjected to whatever physical and/or chemical test is required. In actual practice, such a procedure would be unbelievably time-consuming and in many cases, impossible (e.g., What happens to coated grains? How can cryptocrystalline material be identified and sorted? Can grains as small as 2 μm actually be physically separated and removed from a microscope stage?). Therefore, most modern methods of chemical partitioning attempt to identify the phase and/or the mechanism by which trace elements are associated with sediments without optically examining the samples and picking out individual grains. As such, many of the procedures represent compromises between a conceptual ideal, and practical limitations imposed by time, manpower, and cost. Many of the various methods and their practical uses and limitations are discussed in the following sections.

Table 2.3.4-1. Methods for Determining Chemical Partitioning

- Manual Selection of Phases and Analysis
- Instrumental Selection of Phases and Analysis
- Partial Chemical Extractions
- Density Gradients and Analysis
- Statistical Manipulation of Bulk Chemical Data
- Mathematical Modelling

2.0 PHYSICAL AND CHEMICAL FACTORS AFFECTING SEDIMENT-TRACE ELEMENT CHEMISTRY

2.3 *Chemical factors*

2.3.4 Chemical partitioning methods

2.3.4.1 CHEMICAL PARTITIONING—INSTRUMENTAL METHODS

Relatively new instrumental chemical analytical techniques may be used to determine chemical partitioning in sediments directly (Table 2.3.4.1-1). These methods permit the quantitative analysis of various minerals or mineral assemblages in natural mixtures, in many cases, without prior separation or preconcentration (e.g., without performing magnetic separations, heavy mineral flotations) and sometimes non-destructively from areas as small as 1 μm^2 (Johnson and Maxwell, 1981). Such techniques as x-ray photoelectron spectroscopy (ESCA, XPS) and scanning electron microscopy used in conjunction with energy dispersive x-ray analysis (SEM/EDAX) have been used on many types of geological materials, including sediments, with mixed success (Jones and Bowser, 1978; Johnson and Maxwell, 1981; Horowitz, et al., 1988; Horowitz, et al., 1989d). Many of these procedures permit the user to select the analytical site visually, thus making possible the determination of phase associations for the derived chemical data.

Although these techniques are promising, several problems limit their utility. Foremost is the problem of determining correction factors for the various counting systems used for both the actual analysis and to establish background interference levels. Because of the complexity of these correction factors, detection limits can be as high or higher than 100 mg/kg (100 ppm) (e.g., Johnson and Maxwell, 1981). Further, many of these instruments were designed/developed to be used with manmade materials for purposes of quality control or quality assurance; thus, they usually require that the material under study have an extremely smooth surface. Typical sediment grains, even those having extended contact with water, are simply not smooth enough for use with many of these instruments. Thus, the problems associated with detection limits, as well as the texture of the grains themselves, can limit the usefulness of many instrumental techniques only to major chemical components while excluding environmentally significant trace elements having low concentrations (Table 2.3.4.1-1). On the other hand, if information is needed on a limited number of mineralized components (e.g., examination of sediments affected by mine waste) having well-defined chemistries, then instrumental chemical partitioning procedures can be quite useful (e.g., identification and chemical analysis of arsenopyrite in sediment mixtures showing elevated As levels, Horowitz, et al., 1988; Horowitz, et al, 1989d).

Table 2.3.4.1-1. Instrumental Techniques for Determining Chemical Partitioning

- Electron Microprobe - electron beam producing x-ray emissions
- Scanning Electron Microscope with Energy Dispersive X-ray Analyzer (SEM/EDAX) - electrons producing x-rays
- X-ray Photoelectron Spectroscopy (ESCA, XPS) - photon producing electrons
- U.V. Photoelectron Spectroscopy - photon beam producing ionization and electrons
- Auger Electron Spectroscopy (AES) - electron beam producing ionization and secondary electrons
- Secondary Ion Mass Spectroscopy (SIMS) - positive ion beam producing chemical fragments
- Ion Scattering Spectroscopy (ISS) - ion beam producing primary ions

2.0 PHYSICAL AND CHEMICAL FACTORS AFFECTING SEDIMENT-TRACE ELEMENT CHEMISTRY

2.3 Chemical factors

2.3.4 Chemical partitioning methods

2.3.4.2 CHEMICAL PARTITIONING—PARTIAL EXTRACTION METHODS

One of the oldest and most commonly used methods of chemical partitioning sediments involves the use of partial chemical extractions. The concept of sequential partial extractions is based on the idea that a particular reagent is either phase-specific or mechanism specific (e.g., acetic acid *will only* attack and dissolve carbonates; ammonium acetate at pH 7 *will only* remove adsorbates). Much of the original work in this area was carried out on marine material (Goldberg and Arrhenius, 1958; Hirst and Nicholls, 1958; Arrhenius and Korkish, 1959; Chester, 1965; Lynn and Bonatti, 1965; Chester and Hughes, 1966; 1967; 1969; Chester and Messiah-Hanna, 1970; Cronan and Garrett, 1973; Horowitz, 1974; Horowitz and Cronan, 1976). These early procedures usually involved two-, or at most, three-step sequential extractions plus a total analysis, used in increasing strength, which attempted to chemically partition sediment-associated trace elements. Table 2.3.4.2-1 lists many of the extractants that have been used over the years to partition soils and sediments chemically. Because there are numerous reagents that have been used to selectively solubilize the same geochemical substrate', trace element concentrations ascribed to particular phases and/or mechanisms based on sequential extractions must be viewed as operationally defined.

After the initial work on marine sediments, further applications in chemical partitioning using sequential partial extractions have been made in many diverse fields including: environmental chemistry, pollution studies, water-quality studies, mineral exploration, toxicology, biology etc. (Bruland, et al., 1974; Gupta and Chen, 1975; Brannon, et al., 1976, Chen, et al, 1976; Gambrell, et al, 1977; Luoma and Jenne, 1977a; Malo, 1977; Stoffers, et al., 1977; Jones and Bowser, 1978; Tessier, et al, 1979; Nriagu and Coker, 1980; Forstner and Wittmann, 1981; Diks and Allen, 1983; Chao, 1984; Forstner, 1989; Horowitz, et al., 1989a). These studies used sequential chemical extractions in an attempt to differentiate between anthropogenic and natural trace element pollutants and to try to predict or estimate environmental availability. The tendency was for the sequences to become more and more complex as attempts to clarify chemical partitioning increased. Some procedures involved six, seven, and even nine separate steps (Schmidt, et al., 1975; Brannon, et al., 1976; Summerhayes, et al., 1976; Stoffers, et al., 1977; Forstner, 1982a, b).

Sequential extractions have been used widely on many types of material, but they are not a panacea. The major problem is that the extraction reagents are not as 'selective' as many users assert (Malo, 1977; Pilkington and Warren, 1979; Robbins, et al., 1984; Kheboian and Bauer, 1987). Also, as several investigators have shown, extraction efficiencies vary according to length of treatment and sediment-to-extractant ratio, multiple extractions can solubilize additional concentrations of trace elements, and initial solubilization is counteracted by *in situ* resorption prior to analysis (Malo, 1977; Forstner, 1982a; Robbins, et al., 1984; Gruebel, et al., 1988). However, sequential extraction procedures also have advantages. They permit differentiation between samples having similar bulk chemistries. Also, they represent one of the few practical methods for the determination of concentration mechanisms and thus provide a possible means of estimating bioavailability (Diks and Allen, 1983; Tessier and Campbell, 1987; Campbell and Tessier, 1989). Finally, they offer one of the few potential means of providing information on the concentration of such amorphous materials as reactive iron, amorphous iron oxides, and manganese oxides (e.g., Horowitz and Elrick, 1987; 1989a).

Table 2.3.4.2-1. Selected Reagents Employed in Sequential Partial Extractions

Classification	Reagent
Adsorbates and Exchangeables	0.2M $BaCl_2$-triethanolamine, pH 8.1
	1M NH_4OAc, ph 7
	Distilled Deionized H_2O
	1M NH_4OAc
	1M $MgCl_2$, pH 7
	1M NaOAc, pH 8.2
Carbonates	1M HOAc (25% v/v HOAc)
	1M NaOAc, pH 5 (w/ HOAc)
	CO_2 Treatment
	Exchange Columns
Detrital/Authigenic Hydrogenous/ Lithogenous	EDTA Treatment
	0.1M HCl
	0.3M HCl
Reducible	1M $NH_2OH \cdot HCl$ w/ 25% v/v HOAc
Moderately Reducible (hydrous Fe oxides)	Oxalate Buffer
	Dithionate/Citrate Buffer
Easily Reducible (Mn & amor. oxides)	0.1M $NH_2OH \cdot HCl$ w/ 0.01M HNO_3
Organics	Na Hypochlorite w/ Dithionate/Citrate
	30% H_2O_2 at 95°, pH 2
	30% H_2O_2 w/ 0.02N HNO_3, pH 2 extracted w/ 1M NH_4OAc in 6% HNO_3
	30% H_2O_2 w/ 0.02N HNO_3, pH 2, extracted w/ 0.01M HNO_3
	1:1 Methanol:Benzene
	0.1N NaOH
	0.02M HNO_3 w/ H_2O_2, pH 2, w/ HNO_3 at 85°, w/3.2M NH_4OAc in 20% HNO_3
	30% H_2O_2 in 0.5N HCl, heat
Sulfides	30% H_2O_2 at 95°, extracted w/ 1N NH_4OAc
	0.1N HCl w/ air
Detrital Silicates	$HF/HClO_4/HNO_3$
	Borate Fusion, extracted with HNO_3

2.0 PHYSICAL AND CHEMICAL FACTORS AFFECTING SEDIMENT-TRACE ELEMENT CHEMISTRY

2.3 Chemical factors

2.3.4 Chemical partitioning methods

2.3.4.2 CHEMICAL PARTITIONING—PARTIAL EXTRACTION METHODS

2.3.4.2.1 CHEMICAL PARTITIONING OF SUSPENDED SEDIMENTS BY PARTIAL EXTRACTION

For an example of the type of information that can be obtained using partial chemical extractions, examine the data in Figure 2.3.4.2.1-1. The data were generated by a four-step sequential extraction procedure on suspended sediment collected in the Amazon and Yukon Rivers (Gibbs, 1977). Most of the Cr, Co, and Cu is concentrated in mineral lattices (crystalline); most of the Mn and Ni is associated with ferromanganese coatings. Adsorption was not a significant contributor to the concentration of any of the trace elements studied, and organic matter was a significant concentrator only in the case of Co and Ni.

With these results as a basis, it is possible to discuss potential environmental availability. Trace elements associated with mineral lattices are essentially unavailable. Trace elements associated with ferromanganese coatings probably will be stable or environmentally unavailable unless there is a significant decrease in dissolved oxygen or a significant increase in biological activity. Ferromanganese-associated elements might be released in the digestive systems of certain organisms (e.g., Diks and Allen, 1983). Organics-associated trace elements could be available, particularly if ingested by an organism. Adsorbed trace elements are readily available but, as can be seen, represent a relatively minor percentage of the trace elements present in the suspended sediment.

Solid Phases, in Percent

Phase	Chromium	Manganese	Iron	Cobalt	Nickel	Copper
Absorbed	3.5	0.8	0.007	6.5	3.2	3.5
Coatings	5.2	54.8	43.6	28.8	47.2	5.7
Organics	11.0	6.7	9.0	16.4	15.0	4.5
Crystalline	80.3	37.8	47.4	48.3	34.6	86.2

2.0 PHYSICAL AND CHEMICAL FACTORS AFFECTING SEDIMENT-TRACE ELEMENT CHEMISTRY

2.3 Chemical factors

2.3.4 Chemical partitioning methods

2.3.4.2 CHEMICAL PARTITIONING -- PARTIAL EXTRACTION METHODS

2.3.4.2.2 *CHEMICAL PARTITIONING OF BOTTOM SEDIMENTS BY PARTIAL EXTRACTION*

The data in Table 2.3.4.2.2-1 provide another example of the utility of partial chemical extractions. This information was generated by carrying out a four-step sequential extraction on two marine bottom sediment samples, both having similar Mn concentrations. One sample comes from a hydrothermally active region, and the other sample comes from a hydrothermally inactive region. Ordinarily, hydrothermally active areas display relatively high Mn concentrations. Two questions about these samples need to be addressed: Are the samples geochemically similar, as implied by the bulk chemical results? If they are dissimilar, why?

The acetic acid extraction is designed to remove trace elements held as adsorbates and/or those associated with carbonate minerals. For this extraction, the two sediments are similar. The hydroxylamine extraction is designed to break down manganese oxides. For this extraction, it is obvious that the samples are dissimilar. The sample from the hydrothermally active area has three times as much Mn associated with manganese oxides as has the samples from the inactive area. The hot HCl (hydrochloric acid) extraction is designed to solubilize all but the most resistant silicates and aluminosilicates. Here, too, the sediments are dissimilar; much more Mn is held in this fraction in the sample from the hydrothermally inactive area than in the sample from the active area.

These results show that the sequential partial extractions provided a means of operationally defining how Mn is held in the two sediment samples; they also permitted the differentiation of the two samples. This differentiation would have been impossible solely on the basis of bulk chemical data. Finally, these data re-emphasize the point that sediments which have similar bulk chemistries are not necessarily geochemically similar and that the trace elements associated with them may not have the same level of environmental availability.

Table 2.3.4.2.2-1. An Example of Manganese Partitioning in Bottom Sediment Using Partial Chemical Extractions

| Extraction | Active Hydrothermal Area | | Inactive Hydrothermal Area | |
	Concentration Remaining (mg/kg)	Percentage of Total Removed	Concentration Remaining (mg/kg)	Percentage of Total Removed
Bulk	1,400		1,200	
Acetic Acid	1,300	7	1,100	8
Hydroxylamine	600	50	900	17
Hot HCl	200	29	200	58

2.0 PHYSICAL AND CHEMICAL FACTORS AFFECTING SEDIMENT-TRACE ELEMENT CHEMISTRY

2.3 *Chemical factors*

2.3.4 Chemical partitioning methods

2.3.4.3 CHEMICAL PARTITIONING BY DENSITY GRADIENT AND MINERALOGY

Another method employed to determine chemical partitioning (usually phase rather than mechanistic) uses a combination of physical separations, determination of mineralogy, and subsequent chemical analysis (Muller and Burton, 1965; Francis, et al., 1972; Pilkington and Warren, 1979; Dossis and Warren, 1980). Table 2.3.4.3-1 shows an example of the results for this type of procedure. Each sample was broken down into four relatively broad grain size fractions by settling in acetone (Pilkington and Warren, 1979; Dossis and Warren, 1980). Each size fraction was further subdivided into density subfractions by heavy liquid flotations using various dilutions of tetrabromoethane (bromoform). Each density subfraction was then subjected to mineralogical analysis using a combination of x-ray diffraction and differential thermal analysis techniques. Finally, each density subfraction was chemically analyzed. The data in Table 2.3.4.3-1 are the results from two separate sediment samples taken in the same general area (Spencer Gulf, South Australia). Samples A and D represent the six density subfractions of the 1000- to- 10-μm size separation of each sample. It is apparent that the lightest (least dense) fraction of sample A contains the highest concentrations of the trace elements investigated, while the heaviest (most dense) fraction of sample D contains the highest concentrations. This is true despite the fact that in both cases, these density subfractions constitute only minor components of each sample (8.5 percent for A, and 1.3 percent for D). The D1 subfraction is composed of organic debris, and aggregated particles of magnesian calcite containing varying amounts of quartz, feldspar, mica, kaolin, and goethite while the D6 subfraction is composed of detrital grains of rutile, zircon, ilmenite, tourmalene, magnetite, hematite, and brookite as well as some galena and sphalerite (Dossis and Warren, 1980). Because the mineralogy of each density subfraction is known, establishing chemical partitioning is more direct and certain than with, for example, partial chemical extractions.

This type of procedure also has drawbacks. It is not very useful when there are large percentages of fine particles because of difficulties involved in carrying out density separations on material finer than 10 μm (Pilkington and Warren, 1979). There is no assurance, without individual testing, that the acetone or bromoform will not alter the chemistry [in fact, analyses of the acetone indicate some trace element solubilization (Dossis and Warren, 1980)]. Although knowledge of the mineralogy of each subfraction makes chemical partitioning more informed than, for example, inferring it by using a purely operational definition such as with a partial extraction, the procedure still requires the user to infer partitioning because most density subfractions are mineralogical mixtures. Finally, the method is obviously extremely labor-intensive and consequently, expensive to carry out.

Table 2.3.4.3-1. Density Gradients and Chemical Analysis - Concentration of Trace Elements by Density Subfraction (Data from Dossis and Warren, 1980)

Sample	Density Subfraction	Weight Distrib. (percent)	Zn (µg/g)	Pb (µg/g)	Cd (µg/g)	Zn in Fraction[1] (percent)	Pb in Fraction[1] (percent)	Cd in Fraction[1] (percent)
AS2	D1[2]	8.5	486	436	22.	37	43	38
	D2[3]	6.5	204	166	9.4	12	13	13
	D3[4]	45.7	74	45	1.9	30	24	18
	D4[5]	20.3	85	61	5.4	15	15	23
	D5[6]	0.1						
DS2	D1[2]	2.3	1,580	1,370	25.3	3	4	4
	D2[3]	27.5	1,170	1,140	22.7	13	43	38
	D3[4]	61.6	183	168	1.7	5	14	7
	D4[5]	3.6	648	992	4.1	1	5	1
	D5[6]	3.6	740	1,100	7.3	1	6	2
	D6[7]	1.3	148,000	15,000	607	79	28	49

[1] Percent in fraction calculated by multiplying the concentration by the weight distribution for each subfraction, and then dividing each fractional contribution by the total
[2] D1 subfraction has a density of <2.4 grams per cubic centimeter
[3] D2 subfraction has a density of 2.4 to 2.55 grams per cubic centimeter
[4] D3 subfraction has a density of 2.55 to 2.66 grams per cubic centimeter
[5] D4 subfraction has a density of 2.66 to 2.75 grams per cubic centimeter
[6] D5 subfraction has a density of 2.75 to 2.95 grams per cubic centimeter
[7] D6 subfraction has a density of >2.95 grams per cubic centimeter

2.0 PHYSICAL AND CHEMICAL FACTORS AFFECTING SEDIMENT-TRACE ELEMENT CHEMISTRY

2.3 Chemical factors

2.3.4 Chemical partitioning methods

2.3.4.4 CHEMICAL PARTITIONING USING STATISTICAL MANIPULATION OF DATA

Attempts have been made to determine sediment-chemical partitioning using various statistical manipulations of chemical data. Statistical manipulations may be used to clarify the processes by which various trace elements are partitioned on/in suspended and bottom sediments, especially when dealing with large data sets (e.g., Jones and Bowser, 1978, Horowitz et al., 1989a). Statistical treatments of chemical and/or mineralogical data range from the calculation of relatively simple correlation coefficients through highly complex cluster and multivariate techniques.

For example, Rossman, et al. (1972) used correlation coefficients to partition bulk chemical data for ferromanganese nodules from Green Bay. Because Fe and Mn are the major chemical constituents in the samples, and the Fe and Mn display a strong negative correlation, it is inferred that they form two distinct phases. Additional calculations indicate that the Fe 'phase' appears to concentrate Si and P while the Mn 'phase' appears to concentrate Ba, Ni, Co, Mo, Mg, and Sr (Figure 2.3.4.4-1). Another example, at the other end of the statistical spectrum, is represented by the work of Leinen, et al., (1980), Pisias and Leinen (1980), and Leinen and Pisias (1984). They used a form of factor analysis to partition the chemical data from a large group of sediment samples collected on the Nazca Plate in the Pacific Ocean. The procedure 'identified' five geochemically and mineralogically separate end members (Figure 2.3.4.4-2). A last example is the work of Luoma and Bryan (1981) who applied statistical techniques (linear regressions on log-transformed data) to assess the 'competition' between various geochemical substrates for different trace elements using data from various partial chemical extractions on estuarine bed sediments. Based on this study, geochemical substrate characterization seems best accomplished by Fe and Mn, and humic acid extractions. Further, extractable Fe phases seem more important than total Fe for the concentration of Ag, Cd, Cu, Pb, and Zn, while humic material is also important for Ag and Cu. The results of this study indicate that various substrates 'compete' for different constituents, and that the relative concentrations of differing substrates can strongly influence chemical partitioning.

Statistical manipulations of sediment-chemical data pose some problems because they do not always produce a successful partitioning result. For example, when the Leinen and Pisias (e.g., 1984) procedures were applied to sediments from other areas in the Pacific, the results were much more ambiguous. Another problem with these types of procedures is similar to that encountered with partial extractions: that is, phase identification is indirect. It should be borne in mind that chemical partitioning in both the relatively simple Green Bay study and the more complex Nazca Plate study is based on inference. The fact that a statistical interrelation exists between or among chemical data for a group of samples does not necessarily demonstrate a direct or causal association. At best, and based solely on the statistics, the relations are only inferred, usually on the basis of positive correlation coefficients between the identified 'phases' and their supposed major constituents (e.g., Horowitz, et al., 1989a). Finally, even if various data manipulations produce statistically significant results, they serve no useful purpose if the inferred relations can not be ascribed to a rational environmental setting or process, or if they can not be confirmed by either some other procedure, or a practical test on another data set. Applied statistics should be viewed as an empirical tool used to clarify and evaluate processes or hypotheses; they do not represent an end in themselves (Jones and Bowser, 1978).

Figure 2.3.4.4-1. Statistical Chemical Partitioning of Ferromanganese Nodules from Green Bay, Wisconsin (Data from Rossman, et al., 1972)

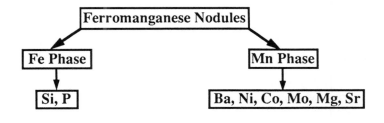

Figure 2.3.4.4-2. Geochemical Endmembers For Sediments From the Nazca Plate, Pacific Ocean (Data from Leinen and Pisias, 1984)

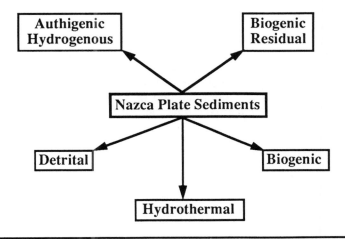

2.0 PHYSICAL AND CHEMICAL FACTORS AFFECTING SEDIMENT-TRACE ELEMENT CHEMISTRY

2.3 Chemical factors

2.3.4 Chemical partitioning methods

2.3.4.5 CHEMICAL PARTITIONING USING MATHEMATICAL MODELLING

According to Luoma and Bryan (1981) the most viable approach to chemically partitioning sediment-associated trace elements may be mathematical models. Others share this view (Vuceta and Morgan, 1978; Leinen, et al., 1980; Swallow and Morel, 1980; Oakley, et al., 1980; Benjamin and Leckie, 1981). The development of mathematical models and their application for determining sediment-chemical partitioning depends on: the availability of adequate thermodynamic and partial extraction data; the ability to identify and quantify geochemical substrates; the development of constants that describe the strength and stability range of trace element-substrate binding, and; the quantification of trace element speciation (solid phase and dissolved) under different physicochemical conditions (Table 2.3.4.5-1).

Numerous laboratory studies have examined the interaction of various trace elements with a variety of geochemical substrates under a wide range of physicochemical conditions. Swallow and Morel (1980) studied the behavior of Cu and Pb in the presence of hydrous iron oxides in artificial seawater under varying pH conditions. The behavior of the iron oxides was considered analogous to a three-dimensional 'sponge' that sorbs trace elements into/onto the solid as they hydrolyze. A model predicting Cu and Pb uptake by iron oxides was derived from the experimental work. Davis and Leckie (1978) studied the uptake of Cu and Ag by iron oxides. They found that complexing ligands affected trace element sorption and either enhanced or suppressed the concentrating capacity of the oxides, depending on whether or not they sorbed to the oxide surface (enhancement by providing additional sorption sites) or remained in solution (suppression by 'competing' with the oxides for available trace elements). These results were apparently confirmed for Cu, Pb, Cd, Co, Ni, and Zn using purely mathematical models (Vuceta and Morgan, 1978; Benjamin and Leckie, 1981) and partial extraction studies (Forstner, 1982a, b).

Using Cu and Cd and a three-part substrate system of clay (bentonite), humic substances (organic matter), and iron and manganese oxides, Oakley, et al. (1980; 1981) tried to develop a mathematical model for predicting chemical partitioning and bioavailability. Laboratory tests of the utility of the bioavailability model using a polychaete worm were somewhat ambiguous; but the authors claimed that substrate-associated trace elements could be more bioavailable than dissolved trace elements at natural concentrations.

Mathematical modelling shows some promise for elucidating the processes involved, and for predicting chemical partitioning in aqueous-sedimentary systems. However, the experiments and calculations on which the models are based were carried out using laboratory generated or purified substrates to eliminate interpretational ambiguities; even in multi-phase systems (e.g., Oakley, et al., 1980), the mixture of substrates are relatively simple compared to the natural environment. Further, there is a lack of adequate data on the identity and structure of many organic constituents and on the stability constants for many trace element-ligand, ligand-substrate, and element-substrate interactions in aquatic systems (e.g., Vuceta and Morgan, 1978). Finally, the results of some modelling studies appear to contradict each other (e.g., Swallow and Morel, 1980 with Vuceta and Morgan, 1978), or the results from more direct analyses or observations (e.g., Swallow and Morel, 1980 with Chester and Hughes, 1966; 1967; Cronan, 1974; Forstner, 1982a; or Davis and Leckie, 1978, Benjamin and Leckie, 1981 with Calmano, et al., 1988; Forstner, 1989).

Table 2.3.4.5-1. Data Requirements for the Construction of Sediment-Chemical Partitioning Models (from Luoma and Davis, 1983)

- Trace Element Binding Intensities
- Geochemical Substrate Binding Capacities
- Concentrations of Various Geochemical Substrates
- Assessment of Geochemical Substrate Binding Capacities When More Than One Substrate is Present (e.g., Organic Coatings on Clay Minerals)
- Assessment of Competition for Binding Sites Between Various Trace Elements and Other Dissolved Species (e.g., Cu vs. Zn vs. Cd and Cu vs. Ca vs. Mg)
- An Understanding of the Reaction Kinetics of Trace Element Redistribution Among Competing Geochemical Substrates

2.0 PHYSICAL AND CHEMICAL FACTORS AFFECTING SEDIMENT-TRACE ELEMENT CHEMISTRY

2.4 The interrelation and relative importance of selected physical and chemical factors affecting sediment-trace element chemistry

At this point we have discussed individually, and in some detail, the major physical and chemical factors which affect the capacity of a sediment to concentrate and retain trace elements (Sections 2.2 and 2.3). In fact, many published studies tend to concentrate only on the effects of one physical or chemical factor (e.g., grain size, surface area, geochemical substrate). However, as pointed out in the introduction to this material (Section 2.1), the individual physical and chemical factors were discussed separately only as a matter of convenience. In actuality, many of these factors are not readily distinguishable from each other, many tend to be synergistic, and some can be competitive (Jones and Bowser, 1978; Forstner and Wittmann, 1981; Salomons and Forstner, 1984; Horowitz and Elrick, 1987). For example, as the mean grain size of a sediment decreases, there are concomitant increases in surface area and the concentrations of many known trace element-concentrating geochemical substrates (Jenne, 1968; Jones and Bowser, 1978; Baker, 1980; Forstner and Wittmann, 1981; de Groot, et al., 1982; Horowitz and Elrick, 1987).

The purpose of the following sections is to attempt to draw together data on the various physical and chemical factors to demonstrate how they interrelate/interact. A major problem associated with such an attempt is the numerous operational definitions that exist for each of the various physical and chemical factors. As pointed out previously, changes in the operational definition of a factor will alter the value of that factor. As a result, much of the data that currently exist in the literature can not be compared or combined. To adequately evaluate the interrelations between the various physical and chemical factors, a consistent data set, based on a single set of operational definitions, must be employed. One of the few such data sets that exist was generated by Horowitz and Elrick (1987) and Horowitz, et al., (1989a). The various operational definitions used in these studies are summarized in Table 2.4-1. The discussions and conclusions that follow are based on that data set; however, it should be understood at the outset that another complete data set, generated by using a different set of operational definitions, might alter the assessment of the relative importance of the various factors or could indicate a different set of interrelations.

Table 2.4-1. Analytical Procedures and Operational Definitions Employed in Quantifying Selected Physical and Chemical Factors (from Horowitz and Elrick, 1987; Horowitz, et al., 1989a)

SAMPLE PREPARATION
   Prescreened By Sieving Through a Non-metallic 2000 μm Sieve
   Drying - Freeze Drying
   Sample Splitting - Coning and Quartering, or a Non-metallic Riffle Splitter
   Sample Grinding - Only for Total Trace Element Analysis, Ceramic Ball Mill
TOTAL TRACE ELEMENT ANALYSIS OF SEDIMENT SAMPLES
Digestion: Hydrofluoric/Perchloric/Nitric Acid in Open Teflon®Beakers (Horowitz and Elrick, 1985)
Analysis: Cu, Zn, Pb, Cr, Ni, Co, Fe, Mn, Al, and Ti - Flame Atomic Absorption Spectrophotometry
      (Horowitz and Elrick, 1985)
         Analysis: As, Sb, Se - Hydride Generation Atomic Absorption Spectrophotometry (Elrick
         and Horowitz, 1986)
Digestion: Hg - Lefort Aqua Regia (Elrick and Horowitz, 1987)
Analysis: Cold Vapor Atomic Absorption Spectrophotometry (Elrick and Horowitz, 1987)
GRAIN SIZE DISTRIBUTIONS
> 63 μm Fractions - Sieving with Non-metallic Sieves on a Shaker Table (Guy, 1969)
< 63 μm Fractions - Air Elutriation (Horowitz and Elrick, 1986)
Mean Grain Size - Graphically from Cumulative Curves (Folk, 1966)
SURFACE AREA
B.E.T. - 30% Nitrogen/Helium Gas Mixture Using a Flowing Gas Single-Point Method
      (Horowitz and Elrick, 1987)
GEOCHEMICAL SUBSTRATE DETERMINATIONS
Total Organic Carbon - Sample Combustion with Infrared Detection of Evolved Carbon
      Dioxide after Pretreatment with 10 Percent Hydrochloric Acid (Plumb, 1981)
Total Organic Matter - Loss on Ignition - Dried Sample Aliquot Combusted in a Muffle
      Furnace at 500°C for a Half Hour (Skougstad, et al., 1979)
Other Organic Matter - Subtraction of Total Organic Carbon from Loss on Ignition (Horowitz and
      Elrick, 1987)
Sequential Extraction Sequence:
Adsorbate/Carbonate Fe and Mn - Sodium Acetate-Acetic Acid Buffer at pH 5 for 5 hours at 25°C
      (Tessier, et al., 1979)
Manganese Oxides/Reactive Fe - Solution of 0.1 M Hydroxylamine Hydrochloride in
      0.1 M HNO_3, Shaken for 30 Minutes at 25°C (Chao, 1984)
Amorphous Fe Oxides and Mn - Solution of 0.25 M Hydroxylamine Hydrochloride in
      0.25 M HCl, Heated and Shaken at 50°C for 30 minutes (Chao and Zhou, 1983)
Organic-Bound Fe and Mn - Solution of 30 Percent H_2O_2 at pH 2 with HNO_3, Heated
      and Occasionally Shaken at 85°C for 5 Hours Followed by Treatment with 3.2 M
Ammonium Acetate in 20 Percent V/V HNO_3 Shaken at 25°C for 30 Minutes
      (Tessier, et al., 1979)
Total Extractable Iron - Summation of All Fe Removed by the Sequential Extraction
      (Horowitz and Elrick, 1987)
Total Extractable Manganese - Summation of All Mn Removed by the Sequential
      Extraction (Horowitz and Elrick, 1987)
Percent Clay Minerals - Semiquantitative X-ray Diffraction Technique (Webster, 1989)

2.0 PHYSICAL AND CHEMICAL FACTORS AFFECTING SEDIMENT-TRACE ELEMENT CHEMISTRY

2.4 The interrelation and relative importance of selected physical and chemical factors affecting sediment-trace element chemistry

2.4.1 Relative importance of physical and chemical factors to sediment-trace element chemistry

     The relative rankings for the various physical and chemical factors discussed in Sections 2.2 and 2.3 are shown in Figure 2.4.1-1 and are based on the hierarchical ranking of a set of correlation coefficients for the various operationally defined (geochemical) factors and the total concentrations of a large group of trace elements. The elements include: Cu, Zn, Pb, Cr, Ni, Co, As, Sb, Se, and Hg. The rankings of the various factors do not apply to each individual element, *per se*, but to the entire group of elements. In other words, the rankings represent a summary of the relative importance of the various factors for the entire group of trace elements. Thus, the relative rankings of the physical and chemical factors for an individual element in the group could differ from the summary rankings.

     The table of relative rankings differs from many similar attempts for several reasons: 1) it combines both physical (e.g., grain size) and chemical (e.g., amorphous iron oxides) factors, 2) it includes surface area which is not usually measured, and 3) it includes several operational definitions of organic matter (LOI, TOC, and OOM). The relative importance of the various physical factors, and the various chemical factors as separate groups do not differ substantially from rankings generated from other studies (Jones and Bowser, 1978; Forstner and Wittmann, 1981; Salomons and Forstner, 1984). The strong ranking for surface area is not terribly surprising considering its inferred interrelation with grain size and the reportedly high surface area values for various geochemical substrates (see Section 2.2.1 and 2.2.8).

     The only major surprise is the relatively low ranking for manganese oxides. Potentially, there are two reasons for this: 1) the operational definition employed for manganese oxides differed substantially from that used by other investigators, and/or 2) the great majority of the samples studied came from freshwater systems where manganese levels were relatively low. This is especially true for comparisons with marine samples which have substantially higher Mn concentrations (e.g., Forstner and Stoffers, 1981; Forstner and Wittmann, 1981).

**Figure 2.4.1-1. Relative Importance of Selected
Physical and Chemical Factors to Sediment-Trace
Element Chemistry (from Horowitz and Elrick, 1987)**

Amorphous Iron Oxides > Surface Area > Total Organic Matter (Loss

on Ignition) > Total Extractable Iron > Other Organic Matter (OOM)

 > Percent Less than 63 μm > Reactive Iron > Total Organic Carbon

(TOC) > Percent Less than 125 μm > Mean Grain Size > Percent Clay

Minerals > Percent Less than 16 μm > Percent Less than 2 μm > Total

Extractable Manganese > Manganese Oxides

2.0 PHYSICAL AND CHEMICAL FACTORS AFFECTING SEDIMENT-TRACE ELEMENT CHEMISTRY

2.4 The interrelation and relative importance of selected physical and chemical factors affecting sediment-trace element chemistry

2.4.1 Relative importance of physical and chemical factors to sediment-trace element chemistry

2.4.1.1 THE INTERRELATION OF GRAIN SIZE AND SURFACE AREA TO EACH OTHER AND TO SEDIMENT-TRACE ELEMENT CHEMISTRY

As pointed out in Sections 2.2.1 and 2.2.4, trace element concentrations tend to increase as grain size decreases (e.g., Jones and Bowser, 1978; Forstner and Wittmann, 1981; Salomons and Forstner, 1984). Table 2.4.1.1-1 contains the correlation coefficients for selected trace elements against a group of grain size ranges commonly used to deal with the grain size effect (Banat, et al., 1972; Cameron, 1974; deGroot, et al., 1982). These data indicate that, in general, the strongest correlations between most of the trace elements and grain size occur for either the percent <125-μm or the percent <63-μm fractions. Correlations with mean grain size (Mz) are also fairly strong (these correlations are negative because as the sediments become finer grained, Mz decreases).

Correlation coefficients calculated for bulk sample chemistry and bulk sample surface area are as strong or stronger than those found for the percent <125- or the percent <63-μm fractions (Horowitz and Elrick, 1987; Table 2.4.1.1-1). Based on these results, it would appear that surface area can be used interchangeably with grain size for normalization purposes to clarify spatial or temporal trends in sediment-trace element chemistry as was done in the Tyrrhenian Sea by Baldi and Bargagli (1982) or for fluvial sediments by Oliver (1973). Data plots of trace element concentrations and surface area, as well as comparisons of the correlation coefficients calculated for log transformed and untransformed data, indicate that the interrelations between trace element concentrations, organic carbon, and grain size, with surface area, are non-linear, possibly logarithmic (e.g., see Figure 2.2.9-1). This is similar to the results reported by Oliver (1973) for surface area with trace elements, and by Suess (1973) for surface area with organic carbon and grain size.

In Section 2.2.8 the surface area of simple spheres was calculated; these calculations indicated that as grain size decreased, surface area increased. If grain size was the major controlling factor for surface area, then the finest fractions should be the major contributors to surface area. If so, correlations between surface area and the various size fractions should increase with decreasing size (e.g., the correlation between surface area and the percent <125-μm fraction should be weaker than between surface area and the percent <63-μm fraction which should be weaker than the percent <16-μm fraction, etc.). However, this does not appear to be the case; the correlations between surface area and the percent <125- and the percent <63-μm fractions are about the same, and both are significantly stronger than between surface area and the percent <16- and the percent <2-μm fractions (see Table 2.4.1.1-1). The surface area measuring technique was supposed to determine only external surface area; therefore, the relation between surface area and grain size should simply be a function of the diameter of the various sediment grains, with departure from ideality (sphericity) being due to a combination of irregular shapes or granular imperfections. Therefore, it would appear that grain size is not the only factor affecting the surface area of sedimentary material. Relative rankings for the importance to sediment-trace element chemistry of the various measures of grain size and surface area are provided in Figure 2.4. 1.1-1.

Table 2.4.1.1-1. Correlation Coefficients[*] for Various Measures of Bulk Sample Grain Size and Surface Area with Bulk Sample Trace Element Concentrations (from Horowitz and Elrick, 1987)

Element	Percent <125 μm	Percent <63 μm	Percent <16 μm	Percent <2 μm	Mean Grain Size	Surface Area
Cu	.91	.93	.53	.51	-.79	.94
Zn	.89	.90	.47	.49	-.78	.92
Pb	.80	.83	-	-	-.82	.85
Cr	.94	.96	.77	.70	-.85	.97
Ni	.78	.84	.70	.61	-.77	.85
As	.72	.81	.75	.73	-.73	.81
Sb	.63	.68	.61	.61	-.58	.76
Se	-	.53	.77	.71	-.67	.66
Hg	.61	.59	-	-	-.53	.69
Fe	.89	.92	-	-	-.73	.93
Mn	.88	.87	-	-	-.65	.89
Al	.88	.90	-	-	-.78	.90
Ti	.88	.87	-	-	-.67	.90
%<125 μm						.93
%<63 μm						.96
%<16 μm						.59
%<2 μm						.56
Mean Grain Size						-.83

[*] - Only correlation coefficients at the 95% confidence interval or higher are presented; therefore, a dash (-) indicates that there was no significant correlation.

Figure 2.4.1.1-1. Relative Importance of Selected Physical Factors to Sediment-Trace Element Chemistry (from Horowitz and Elrick, 1987)

Surface Area > Percent <63 μm > Percent <125 μm >

Mean Grain Size (Mz) > Percent <16 μm > Percent <2 μm

2.0 PHYSICAL AND CHEMICAL FACTORS AFFECTING SEDIMENT-TRACE ELEMENT CHEMISTRY

2.4 *The interrelation and relative importance of selected physical and chemical factors affecting sediment-trace element chemistry*

2.4.1 Relative importance of physical and chemical factors to sediment-trace element chemistry

2.4.1.2 THE INTERRELATION OF GRAIN SIZE AND GEOCHEMICAL SUBSTRATE TO EACH OTHER AND TO SEDIMENT-TRACE ELEMENT CHEMISTRY

The relation between grain size and sediment-trace element chemistry was discussed in the previous section. As pointed out in Section 2.3.2, the concentration of various geochemical substrates increases with decreasing grain size; there are many literature citations which support this contention (e.g., Jones and Bowser, 1978; Baker, 1980; Forstner and Wittmann, 1981; Salomons and Forstner, 1984). In general, the correlation coefficients listed in Table 2.4.1.2-1 for mean grain size (Mz) also support this view. The only major exception are manganese oxides which show no significant correlations with any measure of grain size, and total extractable manganese which shows a limited number. It is interesting to note that the strongest correlations between grain size and geochemical substrate are for the coarser <125- and <63-μm fractions relative to the finer <16- and <2-μm fractions. In fact, in many instances, mean grain size (Mz) displays stronger correlations with the geochemical substrates than do the finer fractions. These results are similar to those for grain size and surface area discussed in the previous section and like the grain size-surface area results, are somewhat unexpected for the same reasons. The relative rankings for the importance of various physical factors to the concentration of various geochemical substrates are provided in Figure 2.4.1.2-1.

Table 2.4.1.2-1. Correlation Coefficients[*] for Various Measures of Bulk Sample Grain Size and Surface Area with Bulk Geochemical Substrate Concentrations (from Horowitz and Elrick, 1987)

Geochemical Substrate	Percent <125 μm	Percent <63 μm	Percent <16 μm	Percent <2 μm	Mean Grain Size	Surface Area
Reactive Iron	.88	.88	-	-	-.74	.91
Manganese Oxides	-	-	-	-	-	-
Iron Oxides	.88	.91	-	-	-.72	.94
Total Extractable Iron	.87	.88	-	-	-.72	.91
Total Extractable Mn[1]	.58	.56	-	-	-	.61
Total Organic Carbon	.91	.92	.83	.66	-.90	.88
Loss on Ignition	.96	.97	.91	.78	-.87	.95
Other Organic Matter	.90	.96	.89	.78	-.87	.94
Percent Clay Minerals	.56	.83	.90	.81	-.76	.68

[1] - Total Extractable Manganese

[*] - Only correlation coefficients at the 95% confidence interval or higher are presented; therefore, a dash (-) indicates that there was no significant correlation.

Figure 2.4.1.2-1. Relative Importance of Selected Physical Factors to Geochemical Substrate Concentration (from Horowitz and Elrick, 1987)

Percent <63 μm > Surface Area > Percent <125 μm >

Percent <16 μm = Mean Grain Size (Mz) > Percent <2 μm

2.0 PHYSICAL AND CHEMICAL FACTORS AFFECTING SEDIMENT-TRACE ELEMENT CHEMISTRY

2.4 The interrelation and relative importance of selected physical and chemical factors affecting sediment-trace element chemistry

2.4.1 Relative importance of physical and chemical factors to sediment-trace element chemistry

2.4.1.3 THE INTERRELATION OF SURFACE AREA AND GEOCHEMICAL SUBSTRATE TO EACH OTHER AND TO SEDIMENT-TRACE ELEMENT CHEMISTRY

The relation between surface area and sediment-trace element chemistry are covered in Section 2.4.1.1. Table 2.4.1.3-1 lists the correlation coefficients determined for surface area and various geochemical substrates as well as with trace element chemistry. Generally, the correlations between surface area and the geochemical substrates are quite high. There are two reasons for this: 1) surface area measurements on the substrates themselves can be high (see Section 2.2.8 and Table 2.2.8-2), and 2) an important factor controlling the concentration and retention of many geochemical substrates (e.g., amorphous iron oxides, organic matter) by sediment grains is surface area. The more surface area a sediment has, the greater its capacity to concentrate various geochemical substrates. Clay mineral percentage, manganese oxides, and total extractable manganese are not correlated strongly with surface area. The lack of a strong correlation between surface area and clay mineral percentage could indicate that the clay minerals are acting as simple surfaces for the deposition of other materials such as iron oxides and organic coatings, and do not contribute directly to overall sediment-surface area (e.g., Jenne, 1976). This view is also supported by the relatively low correlations between clay mineral percentage and trace element concentration.

If the presence of various geochemical substrates contributes to the overall surface area of a sediment, then the removal of selected substrates should produce a concomitant reduction in sediment surface area. This contention was examined during the sequential extraction procedures used to generate this data set (Horowitz and Elrick, 1987). In some cases substrate removal did cause a reduction in surface area; however, in many cases, substrate removal produced a three- to five-fold *increase* in surface area, rather than the expected decrease (Table 2.4.1.3-2). The pattern of change in surface area is grain-size related—in general, sediments having Mzs <125 μm showed increases while those having Mzs >125 μm showed decreases in surface area with substrate removal. This dichotomy in the surface area of sediment was unexpected and requires some explanation (see Section 2.4.2).

The relative rankings for the importance of various geochemical substrates to sediment-trace element concentrations are provided in Figure 2.4.1.3-1 and are based on the correlations listed in Table 2.4.1.3-1. The strong correlations between most of the substrates and sediment-trace element concentrations corroborate much prior work (Jones and Bowser, 1978; Baker, 1980; Forstner and Wittmann, 1981; Salomons and Forstner, 1984). The only unusual result is the lack of a strong relation between trace element concentrations and manganese oxides. However, it is interesting to note the positive correlations between reactive iron and trace element concentrations. The operational definition of reactive iron is the Fe removed during a particular manganese oxide determination (Chao, 1984). The relative importance of manganese oxides was established from two sources: 1) the high trace element concentrations found in manganese nodules (e.g., Mero, 1962) and 2) the release of high trace element concentrations during partial extractions designed to solubilize manganese oxides. At least in the latter case, the importance of manganese oxides, *per se*, may have been overemphasized previously because reactive iron was solubilized simultaneously with manganese oxides.

Table 2.4.1.3-1. Correlation Coefficients[*] for Various Measures of Bulk Geochemical Substrate Concentration and Surface Area with Bulk Sample Trace Element Concentrations (from Horowitz and Elrick, 1987)

Element	Reactive Iron	Manganese Oxides	Iron Oxides	Total Extractable Iron	Total Extractable Manganese	Total Organic Carbon	Loss on Ignition	Other Organic Matter	Percent Clay Minerals
Cu	.88	-	.94	.90	.56	.90	.91	.91	.56
Zn	.91	-	.95	.95	.54	.89	.91	.89	.64
Pb	.77	-	.83	.80	-	.80	.83	.87	-
Cr	.91	-	.92	.92	.55	.94	.99	.97	.71
Ni	.74	-	.84	.80	-	.88	.84	.92	.73
As	.84	-	.87	.87	.54	.76	.84	.78	.82
Sb	.78	-	.79	.74	-	.71	.76	.69	.73
Se	-	-	.51	.64	-	.72	.74	.73	.83
Hg	.73	-	.74	.71	.48	.61	.59	.79	-
Fe	.90	-	.97	.95	.56	.84	.91	.89	-
Mn	.92	-	.93	.91	.82	.72	.85	.70	-
Al	.84	-	.92	.87	-	.85	.85	.90	.55
Ti	.88	-	.91	.88	.48	.83	.87	.79	-
Sur. Area	.91	-	.94	.91	.61	.88	.95	.94	.68

[*] - Only correlation coefficients at the 95% confidence interval or higher are presented; therefore, a dash (-) indicates that there was no significant correlation.

Table 2.4.1.3-2. Effects of Chemical Partitioning and Grain Size on Surface Area (Data from Horowitz and Elrick, 1987)

Extraction	Mean Grain Size (Mz)	Increased Surface Area	Decreased Surface Area
Carbonates and Adsorbates	<63 μm	X	
	>63 μm		X
Manganese Oxides and Reactive Iron	<125 μm	X	
	>125 μm		X
Amorphous Iron Oxides	<250 μm	X	
	>250 μm		X
Organic Matter[*]	<63 μm	X	
	63 - 250 μm		
	>250 μm		X

[*] - Organic matter has three size ranges because it occurs in two distinct forms, coatings (<63 μm and 63-250 μm) and as particles (63 - 250 μm and >250 μm).

Figure 2.4.1.3-1. Relative Importance of Various Geochemical Substrates to Trace Element Concentration (from Horowitz and Elrick, 1987)

Amorphous Iron Oxides > Loss on Ignition (LOI) > Total Extractable

Iron > Other Organic Matter (OOM) > Reactive Iron > Total Organic

Carbon (TOC) > Percent Clay Minerals > Total Extractable Manganese

> Manganese Oxides

2.0 PHYSICAL AND CHEMICAL FACTORS AFFECTING SEDIMENT-TRACE ELEMENT CHEMISTRY

2.4 The interrelation and relative importance of selected physical and chemical factors affecting sediment-trace element chemistry

2.4.2 The interrelation of grain size, surface area, and geochemical substrate to each other

Surface area, grain size, and geochemical substrate are interrelated with sediment-trace element concentrations (Sections 2.4.1.1 and 2.4.1.3). This is in agreement with prior work (Jones and Bowser, 1978; Baker, 1980; Forstner and Wittmann, 1981; Salomons and Forstner, 1984). However, two questions remain to be addressed: 1) if surface area is strongly affected by the diameter of sediment grains, why does it show a strong positive correlation with the percent <125-μm and <63-μm fractions and much lower correlations with the percent <16-μm and <2-μm fractions (Section 2.4.1.1)?; and 2) if the presence of various geochemical substrates contributes to the surface area of a sediment, why does their removal sometimes produce an increase rather than a decrease in surface area (Section 2.4.1.3)? Although these issues are posed as two separate questions, the cause for the observed results is similar.

Two theories could explain the observed changes. The first is based on the view that the geochemical substrates fill in granular imperfections or physical gaps so that their presence produces a smoother, less irregular surface; smoother grains have lower surface areas than rougher ones. This implies that the underlying material has a higher surface area than its coatings. An adjunct to this view entails the disaggregation of agglomerates which are cemented together by the substrates. Thus, the coatings cement fine silt- and clay-sized particles to produce medium- to coarse-silt and very fine sand-sized particles. Removal of the substrate 'cement' breaks down the agglomerates into their original, smaller, individual component particles which have larger surface areas than their aggregates. These infilling/disaggregation theories have been used to explain surface area changes in various soils (Davis, et al., 1971; Curlik, et al., 1974; Lovell, 1975; Egashira, et al., 1977; Baccini, et al., 1982; Kaurichev, et al., 1983). The second theory is based on the view that the extraction procedures alter sediment mineralogy, especially that of the clays, with the new minerals having substantially higher surface areas than their predecessors. Such mineralogical changes are feasible with certain chemical treatments (Heller-Kellai and Rozenson, 1981; Tessier, et al., 1979; Robbins, et al., 1984).

Examination of aliquots of the original material, and the residues after each extraction for a fine- (Mz = 17 μm) and a coarse-grained (Mz= 340 μm) sediment using a combination of x-ray diffraction and SEM/EDAX provided a resolution to the problem. The x-ray diffraction studies indicated that the only detectable mineralogical change was a slight alteration in the hydration of some of the clay minerals (Horowitz and Elrick, 1987). SEM/EDAX examination of the untreated coarse-grained sample indicated that coatings were present and that their surfaces were 'rougher' than the underlying material (Figure 2.4.2-1a); the coatings also appeared to cement finer particles to much larger single grains (Figure 2.4.2-1b). As the extraction sequence proceeded, the size and number of coatings decreased until none were detectable (Figure 2.4.2-1c). Examination of aliquots of the fine-grained sample, from untreated through the extraction sequence, indicated that the sediment became more disaggregated through each step of the sequence. Thus, the aggregates were broken down into uncemented, individual particles (Figures 2.4.2-2a and b). Therefore, although geochemical substrates may make a contribution to sediment surface area, *per se*, this contribution is only important for coarse-grained material, which has a low initial surface area. In the case of fine-grained material, which has a high initial surface area, substrates may increase sediment surface area in their own right, but the effect of grain size overwhelms this contribution. The major effect of substrate removal on fine-grained sediment (Mz ≤ 125 μm) is disaggregation, a decrease in Mz and an increase in surface area.

Figure 2.4.2-1. Scanning Electron Micrographs of Coarse-Grained Sediment from the
Apalachicola River, Florida (from Horowitz and Elrick, 1987)

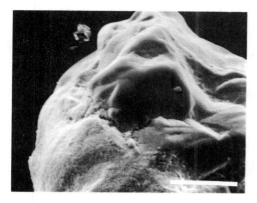

Figure 2.4.2-1a. SEM Micrograph of a Quartz
Grain Partially Covered by an Iron Oxide Coating (Bar = 100 μm)

Figure 2.4.2-1b. SEM Micrograph of Fine Particles
Cemented to a Quartz Grain With an Iron and Manganese Oxide Coating (Bar = 100 μm)

Figure 2.4.2-1c. SEM Micrograph of Quartz Grains After
All Extractions Were Completed (Bar = 100 μm)

Figure 2.4.2-2. Scanning Electron Micrographs of Fine-Grained Sediment
from the Lower Mississippi River, Louisiana (from Horowitz and Elrick, 1987)

Figure 2.4.2-2a. SEM Micrograph of
Untreated Sediment Grains (Bar = 10 μm)

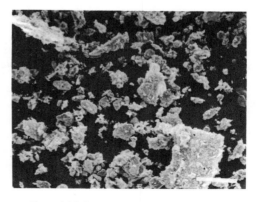

Figure 2.4.2-2b. SEM Micrograph of Sediment
Grains After the Carbonate and Manganese Oxide
Extractions (Bar = 10 μm)

2.0 PHYSICAL AND CHEMICAL FACTORS AFFECTING SEDIMENT-TRACE ELEMENT CHEMISTRY

2.4 The interrelation and relative importance of selected physical and chemical factors affecting sediment-trace element chemistry

2.4.3 Predicting sediment-trace element concentrations using physical and chemical factors

If all the physical and chemical factors described throughout Section 2.0 account for most of the capacity of a sediment to concentrate and retain trace elements, and for much of the observed intersample trace element variability, it should be possible to combine measures of a limited number of these factors to predict sediment-associated trace element concentrations. Predictive models of this type could be superior to those generated from laboratory experimental data because they would be based on highly complex 'real-world' materials. Further, if sample selection was limited to naturally or anthropogenically unaffected areas, the models might provide a measure of normal sediment-trace element concentrations. As such, the models could be used as a reconnaissance tool to identify both potentially affected areas and elevated sediment-associated trace elements (Horowitz, et al., 1989a). Such a set of predictive models for a number of trace elements has been constructed (Table 2.4.3-1). Two models were required for each trace element, depending on the mean grain size of the sediment; one model for materials having mean grain sizes <125 μm and the other model for materials having mean grain sizes > 125 μm.

A reconnaissance survey in a small Louisiana basin provided a test of the utility of this modelling approach in calculating local trace element concentrations. The Mzs for all the Louisiana samples were <125μm; therefore, the <125-μm models were used to attempt to calculate sediment-trace element concentrations Initial comparisons between the predicted and measured trace element concentrations for the samples indicated the need to recalculate the original model coefficients because the calculated concentrations were about double the measured levels. These recalculations should be viewed as a calibration exercise intended to account for local conditions. Calibration was necessary because, although the median trace element concentrations for the Louisiana samples were similar to that of the original samples used to calculate the models, the geochemical factors for the two sets of samples differed significantly. Seven of the revised models were the same (that is, they included all the original independent variables but had different coefficients) or simpler (that is, they included fewer of the original independent variables with different coefficients), while three of the revised models required some alteration of the independent variables relative to the original models. Out of 120 data pairs (measured and model-calculated concentrations) only two (one for Hg, the other for As) did not fall within 10% of each other. The sites of these non-matching data pairs were identified as potentially affected. The elevated Hg level occurred at a site where rice seed traditionally had been treated with mercurial compounds for preservation purposes. This site is also downstream from an oil refinery where Hg compounds are employed as a bactericide. The elevated As level occurred at a site where some urban runoff occurs.

The results from the original modelling exercise, based on 61 samples from a wide variety of hydrologic and geologic environments, as well as the additional results from the Louisiana samples lend support to the view that this modelling technique can be used to establish 'normal' sediment-associated trace element levels. Further, it would also appear to indicate that sediment-associated trace element concentrations and intersample concentration variations can be evaluated using the factors and operational definitions outlined in Section 2.4.

Table 2.4.3-1. Nationwide Models Compared with the Louisiana Sample Models

Element	Source	Model	R	R'2
Cu	NW (<125μm)[1]	$Cu = .674Al + .367LOI + .205Fe \cdot Fe_2O_3 - 1.87$	.93	.84
	NW (>125μm)[2]	$Cu = .520LOI + .351Al + .395Ti + .875$	.93	.85
	LA[3]	$Cu = 1.88Al - .488LOI + .085$	.99	.97
Zn	NW (<125μm)	$Zn = .983Al + .263Fe \cdot MnO_2 + .255LOI + .329$	.94	.87
	NW (>125μm)	$Zn = .548Fe \cdot Fe_2O_3 + .396Al - .300$	.94	.87
	LA	$Zn = 1.867Al + .852Fe \cdot MnO_2 - 1.106LOI - 1.348$	.99	.98
Pb	NW (<125μm)	$Pb = .351SA + .383OOM + .603Ti + .329$	.86	.70
	NW (>125μm)	$Pb = .512Al - .316Ti + .156Mn + .039\%<125\mu m + .756$	.88	.74
	LA	$Pb = 1.384SA - .507OOM - .509Ti - .386$	.99	.98
Cr	NW (<125μm)	$Cr = .580OOM + .442Fe - .151SA + 1.347$	.91	.80
	NW (>125μm)	$Cr = .422Fe \cdot Fe_2O_3 + .419Ti + .166Al + .259$	.89	.77
	LA	$Cr = .499OOM + .320Fe + 1.355$	.99	.99
Ni	NW (<125μm)	$Ni = .963Al + .248TOC + .648$	.96	.91
	NW (>125μm)	$Ni = .387Fe \cdot Fe_2O_3 + .327Al - .266$	.86	.71
	LA	$Ni = 1.652Al - .297TOC + .003$	.97	.92
Co	NW (<125μm)	$Co = 1.661Ti + .089\%<2\mu m + 1.76$	.89	.77
	NW (>125μm)	$Co = .658Fe + .132\Sigma Ex \cdot Mn + .290$	.91	.81
	LA	$Co = .933Fe \cdot Fe_2O_3 - .257\%<2\mu m - .197Mn \cdot MnO_2 - 1.497$	.97	.90
As	NW (<125μm)	$As = 1.282Fe + \%<125\mu m + .690$	.88	.74
	NW (>125μm)	$As = .498SA + .536Fe + .157$	.87	.75
	LA	$As = 2.19Fe - 3.466\%<125\mu m + 6.589$	.99	.96
Sb	NW (<125μm)	$Sb = 1.013Al + .465\%<63\mu m + .145Mn \cdot MnO_2 - 2.062$	.88	.73
	NW (>125μm)	$Sb = .362SA + .346Fe + .518Mz - 1.932$	.86	.71
	LA	$Sb = .402\%<63\mu m - .812$	.95	.90
Se	NW (<125μm)	$Se = .920OOM + .485SA + .616Ti - 1.255$	.88	.74
	NW (>125μm)	$Se = .918LOI - .352Al - .962$	.85	.70
	LA	$Se = 1.196Fe \cdot Fe_2O_3 - 1.195SA + 1.907Ti - 2002$	.98	.94
Hg	NW (<125μm)	$Hg = .605Fe \cdot Fe_2O_3 + .091Mn \cdot MnO_2 - 3.613$	.89	.76
	NW (>125μm)	$Hg = .493Fe \cdot Fe_2O_3 + .224Al - 3.145$	.87	.74
	LA	$Hg = .420TOC + .254\%<2\mu m - 1.408$	.92	.81

NW (<125μm)[1] = nationwide data base models for samples in which Mz<125μm
NW (>125μm)[2] = nationwide data base models for samples in which Mz>125μm
LA[3] = calibrated models for the Louisiana samples, all of which had Mz<125μm

Model Key

Fe = total Fe	Fe·MnO$_2$ = reactive Fe
Mn = total Mn	Fe·Fe$_2$O$_3$ = Fe oxide
Al = total Al	Mn·MnO$_2$ = Mn oxide
Ti = total Ti	ΣEx·Mn = tot. extract. Mn
LOI = loss on ignition	SA = surface area
TOC = tot. organic carbon	Mz = mean grain size
OOM = other organic matter	

3.0 Sediment-trace element data manipulations

3.1 Introduction

The foregoing discussions covering the physical and chemical factors which affect sediment chemistry showed that a variety of properties can alter the distribution and concentration of trace elements associated with both suspended and bottom sediments. Hence, it can be difficult to determine the presence or absence of significant distribution patterns. For example, it may be hard to locate trace element dispersion patterns from a point source of pollution due to dilution with sediments that are coarser and/or less rich in trace elements. However, certain mathematical or graphical manipulations of physical and chemical data may help clarify or determine if significant patterns exist. Various mathematical/graphical manipulations, commonly employed with sediment-chemical data, are listed in Table 3.1-1; they are also described and discussed in the following sections.

Table 3.1-1. Mathematical/Graphical Data Manipulations
Commonly Employed With Sediment-Trace Element Data

Corrections for Grain-Size Differences
Normalization to a Single Grain-Size Range
Carbonate Corrections
Recalculation of Data on a Carbonate-Free Basis
Normalization to Conservative Elements
Multiple Normalizations

3.0 Sediment-Trace Element Data Manipulations

3.2 Limitations of analytical data

Analytical chemists and instrument manufacturers are constantly improving analytical techniques and analytical instrumentation. These improvements have greatly enhanced our ability to detect and quantify lower and lower trace element concentrations. However, none of these improvements have eliminated analytical errors. The evaluation of solid-phase analytical data requires the use of standard materials and practices which permit the calculation of several statistical parameters that provide a measure of the validity and utility of the chemical data. Chemists, geochemists, and others use a series of well-defined procedures and terms in this context. The terms include: reference material, standard reference material, check sample, precision, bias, and accuracy. There are a number of different types of reference materials available from various national and international sources (Flanagan, 1976; Johnson and Maxwell, 1981; Cantillo, 1986; NIST, 1990). The various types of reference materials are listed in Table 3.2-1.

Precision (random error) is the degree of agreement of independent and repeated measurements of the same parameter on the same material. Precision is sometimes referred to as reproducibility. Precision is usually expressed as standard deviation or as relative standard deviation about the mean. The first step in determining the precision of a data set is to calculate the mean of the replicate determinations. Then, the standard deviation and relative standard deviation can be calculated (see Figure 3.2-1). Precision can be determined using any type of material as long as it is homogeneous.

The means, standard deviations, and relative standard deviations cited in Table 3.2-2 were calculated using the formulas given in Figure 3.2-1. Note that the standard deviation for all three reference materials is the same (2mg/kg). However, also note that there are significant differences for the relative standard deviation of each material. As the concentration decreases, the relative standard deviation increases. This is a typical pattern for most analytical results. At or near the detection limit, it is not uncommon for precision to equal the measured concentration.

Bias (systematic error) is the persistent positive or negative deviation of a determined concentration from the assumed or accepted true value. Bias may exist in a single batch of determinations or may occur over a period of time. It should be noted that accurate and precise analytical data can be biased. However, this is only likely to occur if the systematic error is small. Bias is normally determined using reference or certified reference materials.

Accuracy is a function of precision (random error) and bias (systematic error) and measures the nearness of an analytical value to the true value. The accuracy of analytical data only can be evaluated through the use of reference materials. Analysts operationally define or set acceptable limits of accuracy; thus, when an analytical result for a reference material falls within the user's defined confidence limits, the analysis is said to be accurate.

Try to understand and appreciate the limits of laboratory analytical data. At the same time, try to develop a feeling for analytical precision and accuracy. Above all, bear in mind that sampling errors (see Section 4.0) are almost always greater than analytical errors. Therefore, it makes no sense to spend large amounts of time and money carrying out highly precise and accurate analytical procedures if sampling errors are so great as to obviate the utility of such analyses. Finally, do not overinterpret analytical results. For example, if analytical accuracy is ±10 percent and all your results differ by ±5 percent, then to all intents and purposes, the results are the same!

Table 3.2-1. Types of Solid-Phase Reference Materials

Reference Material (RM) - a substance for which one or more properties are sufficiently well-established so it can be used for instrument calibration, the assessment of a measurement technique, or for assigning concentrations to materials.

Certified Reference Material (CRM) - a reference material having values for one or more properties certified by a technically valid procedure accompanied by or traceable to a certificate or other documentation issued by a certifying body.

Standard Reference Material (SRM) - a certified reference material that is prepared and provided by the U.S. National Institute of Standards and Technology (NIST, formerly the U.S. National Bureau of Standards).

Check Standard - a material typically prepared and used 'in- house', to determine analytical precision by repeated analysis over a period of time.

Figure 3.2-1. Formulas for Calculating Various Statistics Used in Evaluating the Quality of Analytical Data

Calculation of Mean Concentration

$$x = \frac{\sum_{i=1}^{n} x_i}{n}$$

where $\overline{x}$ = mean of determinations
x_i = determined value
n = number of determinations

Calculation of Standard Deviation

$$S = \left[\frac{1}{n-1} \sum_{i=1}^{n} (x_i - x)^2 \right]^{1/2}$$

where S = standard deviation
x_i = an individual measurement
$\overline{x}$ = mean of all determinations, and
n = number of determinations

Calculation of Relative Standard Deviation

$$RSD = \frac{S}{x}(100)$$

where RSD = relative standard deviation
S = standard deviation, and
x = mean of determinations

Table 3.2-2. Concentrations and Statistical Parameters for the Determination of Cu in Selected Reference Materials

Determination Number	Estuarine Sediment	USGS SGR-1	River Sediment
1	17	68	105
2	19	63	103
3	20	66	107
4	16	65	109
5	18	67	106
6	20	64	105
Mean	18	66	106
Standard Deviation	±2	±2	±2
Relative Standard Dev.	11.1%	3.0%	1.9%

3.0 SEDIMENT-TRACE ELEMENT DATA MANIPULATIONS

3.3 Corrections for grain-size differences

Figure 3.3-1 has already been discussed (see Section 2.2.4) and shows an obvious relation between grain size and trace element concentration. As pointed out earlier, trace elements tend to concentrate on/in finer-grained sediments. Therefore, the addition of coarser-grained material, which typically has lower trace element concentrations, to a trace element-rich finer-grained material could be viewed as a dilution process. The addition of coarser diluents can readily hide a significant trace element dispersion pattern (deGroot, et al., 1982; Horowitz and Elrick, 1988). One way to deal with the dilution problem is to physically separate out a limited grain size range from a bulk sediment sample and chemically analyze it. However, a reasonably good result may be obtained by determining the percentage of the size range of interest on a separate aliquot of the bulk sample and normalizing the chemical data to it.

The first step is to calculate the dilution factor; the normalized chemical concentration is then calculated by multiplying the bulk sample chemical concentration by the dilution factor:

Dilution Factor = 100/(100 - percent of size fraction greater than the range of interest)

Normalized Chemical Data = (Dilution Factor)(Chemical Concentration in mg/kg)

As an example, assume the grain size range of interest is <63 μm and that the <63-μm fraction constitutes 20 percent of a particular sample; further, the Zn concentration of the bulk sample is 100 mg/kg.

Dilution Factor = 100/100 - 80 = 100/20 = 5

Normalized Chemical Data = (5)(100 mg/kg) = 500 mg/kg Zn

Bear in mind that the underlying assumption for this type of correction is that all, or almost all, of the constituent of interest is found in the size fraction of interest (in this case <63 μm). This is not always true, but if it is, then corrections for grain-size differences may help to clarify a distribution pattern that has been obscured because of varying amounts of diluent (coarser non-trace element-bearing material).

Users of normalization techniques are reminded that data generated in this way are for purposes of clarifying trends and that the data may not reflect true chemical concentrations. The data in Table 3.3-1 clearly support this view, particularly when the fraction or range of interest represents less than 50 percent of the sample. These results indicate that 1) the selected grain-size fraction may not account for all the trace elements in the sample, or 2) there are errors in either the size analysis, the chemical analysis, or both, or 3) the observed intersample chemical differences are not solely the result of grain-size variations. On the other hand, when sample sizes are small, actual physical separations may not be feasible. In this case, mathematical normalization may be the only way to attempt to correct for the grain-size effect.

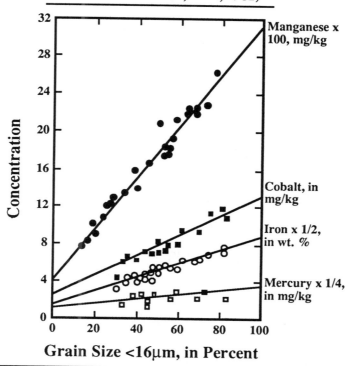

Figure 3.3-1. Interrelation Between Trace Element Concentration and Grain Size for the River Ems, FRG (Data from deGroot, et al., 1982)

Manganese x 100, mg/kg

Cobalt, in mg/kg

Iron x 1/2, in wt. %

Mercury x 1/4, in mg/kg

Concentration (y-axis)

Grain Size <16μm, in Percent (x-axis)

Table 3.3-1. Comparison of Calculated and Measured Trace Element Concentrations for the <63-μm Fraction of Selected Samples (Data from Horowitz and Elrick, 1988)

Sample	Percent <63 μm		mg/kg						wt. percent			
			Cu	Zn	Pb	Ni	Co	Cr	Fe	Mn	Al	Ti
Georges Bank	1	meas[1]	104	55	22	20	12	39	2.0	.02	2.9	.48
M8-5-4		calc[2]	100	500	500	100	107	900	40.0	2.0	20.0	10.0
Columbia Slough	18	meas	47	225	72	31	18	87	4.7	.08	7.5	.68
		calc	145	811	239	145	145	228	20.6	.33	45.0	3.2
Nemadji River	44	meas	31	62	10	31	13	58	3.0	.08	5.8	.40
		calc	50	102	20	48	45	75	5.5	.14	9.8	.86
Patuxent River at	54	meas	35	116	39	32	12	79	3.3	.08	5.6	.42
Point Patience		calc	43	180	46	41	72	89	4.3	.11	7.4	.80
Yaharra River	66	meas	22	45	28	22	8	51	1.9	.06	4.8	.26
		calc	20	41	33	17	27	43	2.0	.08	5.0	.27
Patuxent River at	76	meas	25	131	26	31	14	75	3.5	.06	6.2	.47
Hog Point		calc	26	147	29	36	43	78	4.0	.07	6.5	.62
Georges Bank	89	meas	10	63	22	22	9	70	2.6	.03	5.5	.42
M13A		calc	15	67	22	25	35	62	2.8	.03	5.5	.53
Lake Bruin	95	meas	24	104	23	29	14	65	3.0	.07	7.0	.33
		calc	26	108	23	31	18	53	3.4	.09	6.8	.41

meas[1] - actual determination on an aliquot of the <63-μm fraction

calc[2] - determined by multiplying the bulk chemical concentration of the sample by a normalization factor obtained from the following equation: 100/100 - percent of sample >63 μm

3.0 SEDIMENT-TRACE ELEMENT DATA MANIPULATIONS

3.4 Carbonate corrections

Another commonly employed correction factor, particularly for sediments collected in marine or karstic environments, is the normalization of data to a carbonate-free basis. As with the grain-size correction, the underlying assumption is that the carbonate fraction does not contain substantial quantities of the trace element(s) of interest. The data in Figure 3.4-1, for a marine bottom sediment sample, certainly would support this view, with the possible exception of Pb and Cd, because as Ca increases, the other trace elements decrease [the predominant form of carbonate in marine material is calcium carbonate ($CaCO_3$)]. Also as with the grain-size example, the carbonate fraction might be physically separated and chemically analyzed; however, such a procedure would be very difficult, much more so than for separating a particular size fraction. Corrections of this type have also been used to eliminate the dilution effect of other non-trace element-bearing materials such as quartz, coarse-grained organic matter, etc. (e.g., Thomas, et al., 1972, Horowitz, et al., 1988).

The first step is to calculate the dilution factor; the normalized chemical concentration is then calculated by multiplying the bulk sample chemical concentration by the dilution factor:

Dilution Factor = 100/(100 - percent carbonate)

Normalized Chemical Data = (Dilution Factor)(Chemical Concentration in mg/kg)

For example, assume a sediment sample contains 25 percent carbonate; further, the bulk sample has a Zn concentration of 50 mg/kg.

Dilution Factor = 100/100 - 25 = 100/75 = 1.33

Normalized Chemical Data = (1.33)(50 mg/kg) = 66.5 mg/kg Zn

As an example of how a carbonate correction can clarify a pattern, examine the data in Figure 3.4-2. The broken line represents uncorrected data and the solid line the corrected data. The difference between the two patterns is striking. However, bear in mind that this type of correction assumes that all of the constituents of interest are concentrated in the non-carbonate fraction. This is almost certainly not true for every situation or every trace element, as the data in Figure 3.4-3 show. The pattern for Ca, which represents the dominant form of carbonate ($CaCO_3$), is the same as those for other constituents. Thus, before applying any correction to chemical data, a user must be certain that the underlying assumption upon which the correction is based is valid or, at least, fairly reasonable.

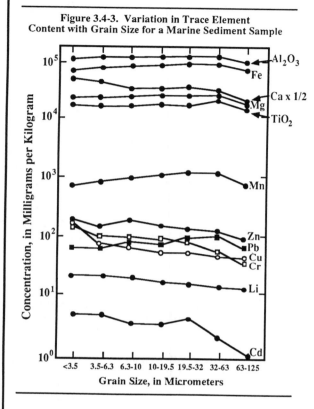

Figure 3.4-1. Variation in Trace Element Content with Grain Size for a Marine Sediment Sample

Figure 3.4-3. Variation in Trace Element Content with Grain Size for a Marine Sediment Sample

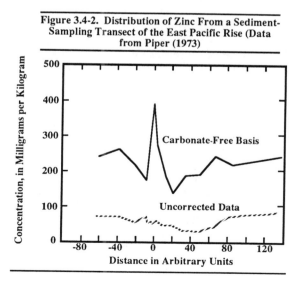

Figure 3.4-2. Distribution of Zinc From a Sediment-Sampling Transect of the East Pacific Rise (Data from Piper (1973)

3.0 SEDIMENT-TRACE ELEMENT DATA MANIPULATIONS

3.5 Normalization to 'conservative' elements

One final commonly employed method of normalizing bulk sediment data entails a calculation with so-called 'conservative' elements. These elements are assumed to have had a uniform flux from crustal-rock sources, from the time sediment particles were eroded until the time they were deposited, over a long period of time; consequently, compensation for changes in levels of various diluents can be made (e.g., Forstner and Wittmann, 1981). The most commonly used conservative elements are Al (Piper, 1973; Bruland, et al., 1974), and Ti (e.g., Forstner and Wittmann, 1981; Horowitz, et al., 1988). Other conservative elements such as Cs and Li have also been used for normalization purposes for the same reasons as Al and Ti (e.g., Loring, 1990). Normalizations of this type entail the determination of a simple ratio:

(Concentration of trace element)/(Concentration of conservative element)

However, unlike the grain-size and carbonate corrections, the resulting values are simply ratios rather than chemical concentrations. This makes comparison with data from other areas difficult, unless the data are similarly recalculated; also, a ratio is more difficult to grasp conceptually than a concentration.

Normalizing to a conservative element can significantly alter, and sometimes clarify distribution patterns, as shown in Figure 3.5-1. The data in that diagram are from a series of marine sediment samples. The broken line represents bulk Cu data corrected to a carbonate-free basis, while the solid line depicts the same bulk Cu data normalized to Al as Al_2O_3. Obviously, the two normalization procedures produce two different patterns. Which one, if either, is correct? The samples come from a transect of the East Pacific Rise, a mid-ocean ridge where the sea floor is actively spreading. The Al-normalization pattern would appear to indicate that the ridge is a source for Cu because it shows a clear-cut peak right over the center (crest) of the ridge. The carbonate-corrected data are not so clear-cut because the Cu peaks occur on the ridge flanks rather than the ridge crest. However, they still might indicate that the ridge is a Cu source, but that the Cu remains in solution or suspension for a period of time before it settles on the ocean floor. Further clarification of the source or processes affecting Cu on active oceanic ridges would require additional sampling and analysis specifically designed to address the issues raised by the two normalized data patterns.

Patterns from normalized data can have the same drawbacks as the statistical manipulation of data (see Section 2.3.4.4). In other words, each pattern must be evaluated in terms of potential trace element sources and potential geochemical processes. If there is sufficient rationale for more than one pattern, then several patterns could be 'correct'; if there only is sufficient rationale to support one pattern, then that pattern alone may be correct. Finally, there may be no process or source rationale for a particular pattern. In such a case, the pattern provides no clarification of environmental processes or trace element sources and simply becomes an unexplained phenomenon. If the rationale for making a normalization in the first place was defensible, and no pattern emerges, this provides useful 'negative' data. For example, in the case of an Al normalization, trace element concentrations have not been affected by changes in input from local crustal rocks.

Figure 3.5-1. Distribution of Copper from a Sediment- Sampling Transect of the East Pacific Rise (Data from Piper, 1973)

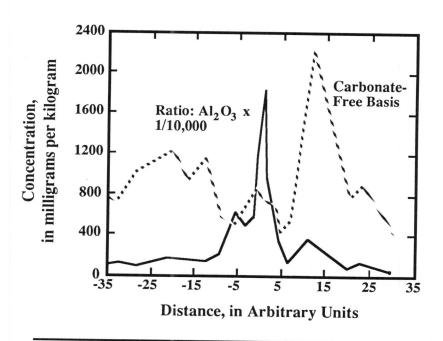

3.0 SEDIMENT-TRACE ELEMENT DATA MANIPULATIONS

3.6 Effects of applying corrections to sediment-trace element data

The preceding sections discussed various types of data normalization techniques and their effect on sediment-trace element distribution patterns. The same normalization techniques also can be applied prior to statistical manipulations or calculations. However, the user should continuously bear in mind that normalization techniques should be applied only with some rationale or hypothesis-testing in mind.

As an example of the effects of data normalization on statistical calculations, examine the correlation coefficients listed in Table 3.6-1. The data come from a core obtained in Lake Oahe, South Dakota. The lake is some 200 miles downstream from Lead, where gold mining and processing have been extant since the 1880s. Concern about the lake stems from the large amounts of As available to some of the rivers feeding it; As is a major constituent of the country rock removed and processed during mining. The questions to be addressed: 1) are there elevated As concentrations in the bed sediments of Lake Oahe?, and 2) if As is elevated, could a source and/or phase be identified?

Elevated As levels occurred in several cores taken in the lake-bed sediments. If arsenopyrite [(FeFsS) the As-bearing mineral in the mined area] was a source for the As, or if some of the As was associated with iron oxides, then there should be some correlation between As and Fe. Correlation coefficients were calculated for the raw (untreated) chemical data; no statistically significant correlation existed for Fe and As. In fact, there were no statistically significant correlations between Fe and any of the trace elements (Table 3.6-1).

Sedimentation in the lake is very rapid (up to 12 cm/year); it is also erratic because it is event-dominated (sedimentation is greater during periods of snowmelt in the spring and following intense thunderstorms in the summer). In such cases, normalization to a conservative element is a fairly typical data manipulation (see Section 3.5). After normalization to Ti, the correlation coefficients were recalculated (Table 3.6-1). Again, there were no significant correlations between Fe and As; only Pb and Co correlated significantly with Fe.

Microscopic examination of the core sections containing the highest As levels indicated that they also contained large amounts of organic matter. Possibly, the As levels were associated with organic matter rather than with Fe. However, there were no significant correlations between organic matter (LOI) and As (Table 3.6-1). The only significant correlations were between LOI and Zn, Co, and Se, and they were all negative, implying that the organic matter might be acting as a diluent, rather than as a concentrator/collector. If the organic matter was acting as a diluent, then normalization of the data to organic matter should eliminate the dilution effect. Recalculation of the correlation coefficients after normalization to organic matter produced surprisingly high correlation coefficients for most of the elements and Fe, but not for Fe and As (Table 3.6-1). These results would indicate that organic matter was acting as a diluent for most of the trace elements in the core samples.

The source of the organic matter is almost certainly local (from within the lake itself) because the areas around the rivers feeding the lake are sparsely vegetated. Further, the organic content of the river sediments is low. On the other hand, sedimentation in the lake, (see above), can be rapid and erratic. The source of the majority of the As is almost certainly not local (i.e., it is probably coming from the bed, banks, and floodplains of Whitewood Creek and the Belle Fourche River which flow into the Cheyenne River which flows into the lake). Thus, two factors had to be considered: the erratic sedimentation, and the dilution effect of the organic matter. Therefore, the data were normalized to both Ti (to deal with the sedimentation) and to LOI (to deal with the dilution) and the correlation coefficients were recalculated (Table 3.6-1). Now, all the trace elements, including As, correlate strongly with Fe. Bear in mind that these final results *do not prove* the presence of arsenopyrite, or As associated with iron oxides, but do indicate the processes going on in the lake. The presence or absence of As-bearing phases still must be confirmed by other means; in this study this was done using SEM/EDAX (Horowitz, et al., 1988).

Table 3.6-1. Correlation Coefficients[*] for Raw and Variously Normalized Fe and Selected Trace Element Concentrations and for Loss on Ignition (LOI) and Selected Trace Elements From a Core Obtained from Lake Oahe, South Dakota (Data from Horowitz, et al., 1988)

Element	Raw Data	Normalized to Ti	Correlated with LOI	Normalized to LOI	Normalized to Ti/LOI
Cu	-	-	-	.95	.94
Zn	-	-	-.66	.97	.94
Pb	-	.56	-	.64	.89
Ni	-	-	-	.94	.84
Co	-	-.62	-.48	.83	.74
Cr	-	-	-	.93	.83
As	-	-	-	-	.81
Sb	-	-	-	.85	.91
Se	-	-	-.49	.87	.53
Hg	-	-	-	.47	.60

[*] - Only correlation coefficients significant at the 95 percent confidence limit, or higher, are reported; therefore, a dash (-) indicates there were no significant correlations.

91

4.0 SAMPLING SEDIMENTS

4.1 Sampling sediments—general considerations

Sampling and program design are extremely important to the ultimate success of any study. More time, effort, and money have been wasted in environmental studies because of poor program design than for any other reason (Keith, et al., 1983; Hakanson, 1984; Keith, 1988). Further, it is fairly safe to say, although statistical proof is lacking, that more errors are introduced into a study by improper sampling design and sample handling (these are, in fact, two separate issues, with only the former amenable to statistical analysis; Gy, 1979) than could ever be attributed to analytical errors introduced in a laboratory.

The decisions on how, when, and where to sample must be predicated on what an investigator is trying to discover and how the data are to be used. Also, a thorough knowledge of local hydrologic (e.g., flow regimes) and geologic conditions (e.g., information on local rock types) can substantially simplify program planning. In addition to obtaining knowledge about local conditions, it is a good idea to obtain both statistical and analytical chemical advice during program and sampling plan development if a project chief lacks expertise in these areas (e.g., Keith, 1988).

Numerous books have been written solely on the subject of environmental program planning and sampling design (e.g., Watterson and Theobald, 1979; Green, 1979; Gy, 1979; Sanders, et al., 1983; EPRI, 1985; Keith, 1988). This Primer is not the place for an exhaustive discussion of environmental program design and/or sampling for various types of environmental studies; however, Table 4.1-1 provides some information on four types of very commonly employed sampling programs and their limitations. In addition, Table 4.1-2 contains a series of practical suggestions that all environmental program planners should address during the design phase of a study (Green, 1979).

The final subject in Table 4.1-2 requires special mention. Program planning and design does and should not end with the start of actual sampling and analysis. Planning and design should be viewed as an ongoing process. It is a 'little' late, for example, at the end of a two-year study to decide that the initial sampling sites were poorly located, or that samples taken in the first year require reanalysis. In other words, all aspects of a study should be reevaluated periodically in light of meeting program goals and acceptable statistical confidence limits. If they are not being met, redesign may be required.

The following sections provide some practical information specific to sediment (both bottom and suspended) sampling. They are based on actual experiences in the field and in the laboratory. More detailed discussions of the various topics can be found in the cited references in the appropriate sections and in others listed in the Reference section of this Primer.

Table 4.1-1. Examples of Commonly Employed Sampling Designs

<u>Simple Random Sampling</u>—sites are selected such that every possible sample has an equal chance of being chosen. This method is very efficient in homogeneous areas; however, it may be ineffective in heterogeneous areas and could lead to overlooking important sites or data. This method is commonly used in reconnaissance surveys where little is known about local conditions.

<u>Stratified Random Sampling</u>—entails dividing a heterogeneous area into homogeneous subareas within which sampling locations are randomly selected. This often permits the elucidation of subtle but real differences. This method requires some knowledge of local conditions.

<u>Systematic Sampling</u>—entails establishing a constant interval of distance between sampling sites determined by the number of planned samples. Randomness is introduced through the selection of the initial sampling site. The advantage of this method is its ease of application; the disadvantage is that it can produce very biased results.

<u>Fixed Transect</u>—sampling occurs at fixed and predetermined sites that need not be at constant intervals. This method, while extremely simple, has a major statistical drawback; since sites do not have an equal chance of being selected, any inferences or conclusions are associated only with the selected sites. Thus, areal conclusions may not be valid.

Table 4.1-2. Practical Subjects that Environmental Program Planners Should Address (from Green, 1979)

• Replicate sampling for controlled variables

• Equal numbers of randomly allocated replicates

• Collection of samples in the presence and absence of various environmental conditions which may affect the subject of the study

• Initial reconnaissance surveys to establish local conditions and to identify homogeneous subareas within a heterogeneous area

• Verification of the efficacy of the sampling device(s) over the range of conditions likely to be encountered during the study

• Ongoing statistical analyses as data is generated to assess the initial sampling design as it affects such factors as number of sites, location of sites, frequency of sampling, physical size of the sample, etc.

• Ongoing quality control of data as it is collected to permit timely analytical reruns and/or resampling for verification of anomalous values

4.0 SAMPLING SEDIMENTS

4.2 Bottom sediments

Bottom sediments can be used for a variety of studies (Table 4.2-1). Generally, bottom sediment sampling is divided into two distinct categories: surficial sampling and sampling at depth. In certain cases, either sample type may be appropriate depending on program goals. Surficial samples typically are used for the determination of trace element-spatial distributions and reconnaissance surveys and are generally limited to the collection of only the upper 1 to 3 cm of the sediment column. Sampling sediments at depth typically is used to: determine temporal trace element-trends, establish local trace element-baseline levels, and/or establish historical changes in sediment-associated trace element concentrations.

Many different devices have been designed and used over the years to obtain these types of sediments in a variety of environmental settings. Bed sediment samplers fall into three broad classifications: 1) grab samplers, 2) corers, and 3) dredges. Corers generally collect both surficial and sediment column samples and show the least amount of disturbance; grabs collect large surficial samples; and dredges collect even larger, well-mixed near-surface samples. Usually, dredge samples are considered to be qualitative because their use does not permit adequate control of sample location or sampling depth in the sediment column (Plumb, 1981).

Table 4.2-1. Various Types of Trace Element Studies in Which Bed Sediment Samples Have Been Used

- Establishing Geochemical/Trace Element Cycles
- Water Quality Reconnaissance Surveys
- Geochemical Exploration Reconnaissance Surveys
- Regional Geochemical Surveys
- Determining Trace Element Spatial Distributions
- Determining Long-Term Temporal Changes in Trace Element Concentrations
- Establishing Local Baseline Concentrations
- Identification of Point and Non-Point Sources of Pollution
- Monitoring Aquatic Disposal of Wastes
- Determining Biological Effects

4.0 SAMPLING SEDIMENTS

4.2 Bottom sediments

4.2.1 Sampling surficial bed sediment

There is ample evidence from numerous geochemical exploration surveys (e.g., Hawkes and Webb, 1962; Watterson and Theobald, 1979) and from regional geochemical reconnaissance studies (e.g., Webb, 1978; Fauth, et al., 1985) that surficial bed sediments can provide an excellent synoptic picture of trace element-spatial distributions. Typically, such surveys entail random sampling over large geographical areas using stream sediments collected from small, localized streams.

In the case of shallow, wadeable streams, samples are usually collected by hand; in the case of deeper rivers, ponds, or lakes, samples are usually collected with some type of grab sampler (Sly, 1969; Plumb, 1981; OWDC, 1982; Norris, 1988). There are numerous grab sampling devices, of various design, that have different advantages and disadvantages depending on the nature of the sediment to be sampled (e.g., coarse versus fine), the water depth, the amount (mass) of sediment required, the size of the area to be sampled, local energy conditions (e.g., sampling in a rapidly flowing stream versus sampling in a relatively quiescent lake), sampling platform (e.g., a boat versus sampling from a bridge), the availability of lifting equipment (e.g., hand-operated versus crane- or winch-operated), etc. Table 4.2.1-1 lists a large variety of grab samplers with some limited comments about their applicability and utility. Generally, the selection of a particular type of grab for the collection of a sediment-trace element sample is dependent on evaluations against four criteria: 1) degree of physical disturbance during sampling, especially while the device is being lowered to collect a sample (due to the 'bow or pressure wave' created by the device which can disperse fine-grained sediment or flocs at the sediment-water interface); 2) loss of material, especially fine-grained sediments, during recovery of the sampler through the water column ('washout'); 3) the efficiency of the grab for collecting sediments of varying textures (e.g., grain size, degree of induration); and 4) potential for sample contamination.

The evaluation of a number of different grabs, using these criteria, has led some investigators to the conclusion that the best type of surficial-sediment sampler is a coring device (e.g., Norris, 1988). However, corers typically are not used for areal surveys based on surficial sediment samples, especially in shallow, wadeable aquatic environments (see Section 4.2.2). This is because a major disadvantage of most corers is the extremely small area of the bed that is actually sampled. Thus, many more core samples than grab samples usually are required to provide an adequate bottom sediment sample.

One of the most important considerations when collecting surficial sediments is that of obtaining a representative sample. There are various statistical procedures for addressing this problem, depending on the degree of confidence required to meet particular program goals (Keith, et al, 1983; Hakanson, 1984). The confidence limit is also affected by the number of samples to be collected in a particular study area, how the data are to be used, and the degree of geochemical detail required. As a result of all these factors, regardless of the requisite degree of confidence, it is invariably better to collect a group of sub-samples to generate a final composite sample than to arbitrarily collect a single isolated sample as being representative of a sampling site.

Table 4.2.1-1. Partial Listing of Commonly Used Grab Sampling Devices and the Types of Sediments Where They Are Most Effective (Data from Sly, 1969)

- **Dietz-LaFond-Grab**: small samples of soft clay, mud, silt, sand

- **Birge-Ekman Grab**: small and bulk samples of soft clay, mud, silt, silty-sand

- **U.S. BMH-60 (rotating bucket) Grab**: small samples of soft clay, mud, silt, and silty-sand

- **U.S. BM-54 (rotating bucket) Grab**: small samples of soft clay, mud, silt, and silty-sand

- **Franklin-Anderson Grab**:bulk samples of soft clay, mud, silt, sand, rarely gravel

- **Petersen Grab**: bulk samples of soft clay, mud, silt, sand, gravel

- **Ponar Grab**: bulk samples of indurated or soft clay, mud, silt, sand, gravel

- **Shipek Grab**: bulk samples of indurated or soft clay, mud, silt, sand, gravel

- **Van Veen Grab**: bulk samples of indurated or soft clay, mud, silt, sand, gravel

- **Smith-McIntyre Grab**: bulk samples of indurated or soft clay, mud, silt, sand, gravel

- **Orange Peel Bucket Grab**: bulk samples of indurated or soft clay, mud, sand

4.0 SAMPLING SEDIMENTS

4.2 Bottom sediments

4.2.2 Sampling bed sediments at depth

Vertical sampling of a sediment column invariably involves the use of some type of coring device. These tend to fall into three major categories: 1) gravity corers, 2) piston corers, and 3) vibrocorers (Table 4.2.2-1). Many of the criteria that apply to the selection of a grab sampler also apply to the selection of a coring device (see previous section). One additional criteria is the length of sediment column to be sampled. Selection of core samples invariably involves subsampling, especially when there are obvious physical differences (e.g., texture, color) between various sections of an entire core .

Gravity corers, as the name implies, use the force of gravity to penetrate into the sediment column and obtain a sample. Generally, the heavier the corer, the greater the degree of penetration. These devices also require a minimum amount of water depth to achieve sufficient velocity to obtain maximum penetration. To some extent, the amount of weight required can be counterbalanced by the thickness of the core barrel (the thinner the barrel, the lower the resistance to penetration), and by reducing the degree of water resistance to the speed of descent (larger diameter barrels produce less resistance; also, the type of valve at the top of the corer, usually required to prevent sample loss during recovery, can affect the degree of water resistance). Box corers and 'Kastenlots' are special types of 'gravity' corers which do not require rapid rates of descent to deeply penetrate a sediment column, . However, both devices are usually very heavy. Box corers scoop out a section of the sediment column through the operation of a set of springs which are triggered after the device is lowered to the sediment bed. Kastenlots are extremely heavy and wide barrelled, with the barrel walls being made of extremely thin but very rigid material. These devices are slowly lowered to the sediment bed and achieve high levels of penetration because of their weight working in combination with their lack of frictional resistance due to the thin walls of their barrels. Typical gravity cores do not exceed 6 feet in length (~1.8 meters) although Kastenlot cores of up to 20 feet (~6.1 meters) have been recovered.

Piston corers are used to obtain long cores in relatively soft sediments. They are usually very heavy. They are set up so that the piston, which is inserted inside the barrel, stops at the sediment-water interface while the core barrel continues to penetrate the sediment column. The piston creates a vacuum which reduces frictional resistance to barrel penetration. Under the right conditions, piston cores of more than 100 feet in length (~30.5 meters) have been collected.

Long cores in fairly indurated sediments are normally obtained with a vibrocorer. These devices can be powered with either electricity or compressed air. Sediment sampling is achieved through the use of thin-walled barrels in conjunction with vibration which tends to 'fluidize' the sediments to facilitate penetration. As a result, vibrocores tend to be more disturbed than piston cores. Vibrocore length is controlled by the size of the system being used, but typically, does not exceed 40 feet (~12 meters).

Table 4.2.2-1. Partial Listing of Commonly Used Core Sampling Devices

- <u>Free-Fall Gravity Corers</u>: used without a wire or line for short-length cores in very soft fine-grained sediment

- <u>Standard Gravity Corers</u> : used with a wire or line for short- to medium- length cores in soft fine-grained clay to fine-grained sand

- <u>Box Corers</u>: used with a wire or line for short-length undisturbed cores with a large surface area in soft fine-grained clay to coarse-grained sand

- <u>Kastenlots</u>: used with a wire for medium-length undisturbed cores in soft fine-grained clay to medium-grained sand

- <u>Piston Cores</u>: used with a wire for medium- to very long-length relatively undisturbed cores in soft fine-grained clay to medium-grained sand

- <u>Vibrocores</u>: used with a wire for medium- to long-length cores in soft fine-grained clay to indurated coarse-grained sand

4.0 SAMPLING SEDIMENTS

4.3 Suspended sediments

Suspended sediment can be used for a variety of trace element geochemical and water-quality studies (Table 4.3-1). This material plays an extremely important role in the biological and geochemical cycling of trace elements in fluvial systems (e.g., Forstner and Wittmann, 1981; Forstner and Salomons, 1984). As pointed out earlier, suspended sediment-associated trace elements may account for the majority of the transport of trace elements from the continents to the oceans (see Sections 1.3 to 1.6). Thus, the sampling and analysis of suspended sediments is a requisite for any studies involving the determination of trace element transport and the calculation of trace element fluxes. In addition, suspended sediment, along with the sampling and analysis of dissolved samples, may represent the only available means of determining short-term temporal changes in water quality.

Suspended sediment transport is strongly interrelated to both hydrological and geomorphological characteristics (OWDC, 1982; Salomons and Forstner, 1984). The major interactive processes which affect sediment transport include: 1) erosion from floodplains, banks, and channels (availability); 2) vertical particle movement in the transporting medium (lifting capacity, affect of turbulence); 3) horizontal movement by streamflow (stream velocity or discharge); 4) deposition in the channel; and 5) compaction of sediments after deposition (Raudkivi, 1967; OWDC, 1982; Salomons and Forstner, 1984). As a general rule, assuming enough material is available, as fluvial discharge or velocity increases, suspended sediment concentrations also increase.

Grain size plays an extremely important role in the transport behavior of sediment particles (Hjulstrom, 1935; Vanoni, 1977; OWDC, 1982; Salomons and Forstner, 1984). Silt and clay (material <63 μm) are readily suspended and travel at about the same velocity as the water (Guy, 1966). However, the inception of motion of fine particles requires substantially higher stream velocities than would be expected solely on the basis of grain size. This is due to the cohesiveness of fine sediments; a factor which tends to resist the process of resuspension (Hjulstrom, 1935; Vanoni, 1977). However, once suspended, fine sediments tend to remain in suspension for long periods of time and can traverse long distances. Further, although there is much controversy on this point, some researchers believe that the <63-μm material, once suspended, is homogeneously distributed in fluvial cross-sections (e.g., Vanoni, 1977; Ongley and Blachford, 1982).

Sand-sized and coarser particles (>63 μm) are found in many stream channels and are usually transported both in suspension and along the stream bed (bedload). The type and rate of movement of this coarser material is controlled by the the size of the particles, their density, the velocity of the transporting medium, and channel geometry (Vanoni, 1977; OWDC 1982; Salomons and Forstner, 1984). Under normal flow regimes, the bedload consists of only a relatively thin layer of material just above the stream channel; this can change substantially when flow changes from laminar to turbulent (Salomons and Forstner, 1984). However, even under fairly turbulent conditions, material coarser than 63 μm is rarely homogeneously distributed, either horizontally or vertically in fluvial cross-sections (e.g., Guy, 1966; OWDC, 1982). Coarse particle movement tends to be of relatively short duration, interrupted by periods of quiescence when the particle again settles on/in the bed (OWDC, 1982).

Table 4.3-1. Various Types of Trace Element Studies in Which Suspended Sediment Samples Have Been Used

- Determining Geochemical/Trace Element Cycles
- Water Quality Reconnaissance Surveys
- Geochemical Exploration Reconnaissance Surveys
- Regional Geochemical Surveys
- Determining Trace Element Spatial Distributions
- Long- and Short-Term Temporal Changes in Trace Element Concentrations
- Establishing Local Baseline Concentrations
- Identification of Point and Non-Point Sources of Pollution
- Monitoring Aquatic Disposal of Wastes
- Determining Biological Effects
- Determining Trace Element Transport
- Establishing Trace Element Fluxes

4.0 SAMPLING SEDIMENTS

4.3 Suspended sediments

4.3.1 Sampling suspended sediment in fluvial environments

In water-quality studies, as in other studies, the collection of a representative sample is of extreme importance as it is almost impossible to adequately sample and analyze an entire water body (e.g., Horowitz, et al., 1989c). Suspended sediment samplers fall into three general categories: 1) integrating samplers which accumulate a water-sediment mixture over time; 2) instantaneous samplers which trap a volume of whole water by sealing the ends of a flow-through chamber; and 3) pumping samplers which collect a whole-water sample by pump action (OWDC, 1982). Integrating samplers are usually preferred because they appear to obtain the most representative fluvial cross-sectional samples (Table 4.3.1-1) (e.g., OWDC, 1982).

Representative sampling in fluvial cross-sections to determine suspended-sediment concentrations and to subsequently quantify associated trace elements has long been a subject of controversy (Table 4.3.1-2). One view suggests that representative sampling of suspended sediment requires a composite of a series of depth- and width- or point-integrated, isokinetic samples obtained either at equal-discharge or at equal-width increments across a river (Feltz and Culbertson, 1972; Vanoni, 1977; OWDC, 1982, ASTM, 1990). Another view is that only the <63-μm fraction of suspended sediment is important for geochemical and/or water-quality studies. Because this material is believed to be evenly distributed in fluvial cross-sections, a surface or near-surface 'grab' sample taken at or near the centroid of flow will provide a representative sample (e.g., Ongley and Blachford, 1982). Additional suggestions indicate that the depth of a stream, or its discharge should control the type of sampling procedure employed (OWDC, 1982).

Most sampling equipment and sampling designs are established to obtain an 'instantaneous' representative sample (e.g., OWDC, 1982). However, there is substantial evidence to indicate that temporal changes in suspended sediment concentration and cross-sectional distributions can be quite large and therefore, samples should be obtained over a long period of time to be truly representative (e.g., for 8 to 10 hours; Ongley and Blachford, 1982; Horowitz, et al., 1989c; Horowitz, et al., 1990). Unfortunately, no single sampling device, nor technique, simultaneously deals with both cross-sectional (spatial) variability and temporal variability. The user must decide which variable is more important to a study, and must select a sampler and technique accordingly.

The issue of spatial versus temporal variability is of particular importance if a study is designed to address the issue of annual transport or trace element fluxes. This problem is further exacerbated by manpower and logistical problems associated with specific event-type sampling. For example, there is an old adage about fluvial transport which says "90 percent of the transport can take place during 10 percent of the time". If fluvial transport follows this maxim, then temporal variability is of paramount importance for the determination of annual transport and the sampler and sampling design must be tailored to assess temporal variability. Unless a sampling crew can be available 24 hours a day, 365 days a year (normally beyond the financial and manpower resources of most programs) this implies that some type of automatic sampler is required. The tradeoff is that almost all automatic samplers are incapable of obtaining a depth-integrated, isokinetic, and spatially representative sample. Thus, the typical pragmatic solution to this problem is a compromise involving the collection of both manual and automatic samples; the final mixture being predicated on the characteristics of the site(s) under study.

The following sections provide examples of the types of differences that can occur due to the use of different types of samplers or sampling designs. They also provide data on the magnitude of spatial and temporal variability in suspended sediment and associated trace element concentrations that can be found when working with suspended sediment.

Table 4.3.1-1. Requirements for a Suspended Sediment Sampler (from U.S. Interagency Report, 1940)

- Collected sample must be representative of the water-sediment mixture in proximity to the sampler intake at the time of sampling
- The sampler must not alter the concentration of sediment in the whole water sample during collection
- Sampler volume must be sufficient to satisfy the laboratory requirements for all subsequent physical and chemical analytical work
- Sampler must be usable in streams of any depth and for sampling at a particular depth from the surface to the bottom
- Sampler should should be streamlined and of sufficient weight to reduce deflection from the vertical or horizontal planes even under high velocity conditions
- Sampler design should be as simple as possible to limit cost and maintenance
- Sampler should be constructed of non-contaminating material(s) for any requisite analytical work

Table 4.3.1-2. Types of Fluvial-Suspended Sediment Integrating Sampling Procedures (from OWDC, 1982)

Point Integration: The flow area is divided into lateral increments with samples collected isokinetically at various depths along a vertical in each increment; incremental widths and sampling intervals are selected such that concentration and velocity differences between adjacent points are sufficiently small as to meet desired accuracy.

Depth Integration: An isokinetic sample is collected while the sampler is moved vertically at a uniform speed; sampling can be continuous or discontinuous over the entire depth.

Equal Discharge Increment Vertical (EDIV): One of several vertical depth integrated isokinetic samples collected at the centroid of an equal flow segment in a riverine cross section.

Equal Discharge Increment Sample (EDI): A suspended sediment collection technique designed to obtain an isokinetic discharge-weighted sample at a riverine transect by (a) performing vertical depth integration at the centroids of equal-flow segments across the transect and by (b) using a vertical transect rate at each vertical that provides equal sample volumes from each flow segment.

Equal Width Increment Sample (EWI): A suspended sediment collection technique designed to obtain an isokinetic discharge weighted sample at a riverine transect by (a) performing vertical depth integration at a series of vertical sites equally spaced across the transect, and by (b) using the same vertical transit rate at all sampling verticals.

Composite Sample: An actual sample, formed by combining collected EDIV samples, which is representative of the vertical and horizontal distribution of suspended sediment in a riverine cross section.

4.0 SAMPLING SEDIMENTS

4.3 Suspended sediments

4.3.2 Cross-sectional spatial and temporal variations in suspended sediment and associated trace elements and their causes

A major evaluation of sampling design, sampler comparisons, and spatial and temporal variability in suspended sediment and associated trace element concentrations was carried out in 1987 by the U.S. Geological Survey (Horowitz, et al., 1989c; Horowitz, et al., 1990). The data provided in the following sections are all excerpted from that study, and represent one of the few complete examples of the differences due to sampler selection, and the magnitude of spatial and temporal variability in suspended sediment and associated trace element concentrations. The study was carried out at 6 different sites around the U.S. on a variety of streams. With the exception of one instance, all samples were obtained during periods of constant discharge.

At each site, mean velocity was determined using a current meter. Water discharge was computed and five equal discharge incremental vertical station locations (EDIV) were identified. In addition to the five EDIV stations, a location near the centroid of flow was selected for the placement of a point sampler and a pumping sampler. Both point and pump samples were collected at 20 percent of depth measured from the surface (Figure 4.3.2-1). The length of depth-integrated sample collection was limited to produce sample volumes on the order of 2500 to 2700 mL (sampler capacity was 3000 mL). Both the point and pump samples were obtained contemporaneously with, and for the same period of time as, the depth-integrated samples.

The first set of EDIV samples were collected simultaneously at each sampling site by five two-man crews and were retained for later compositing. This composite provided an initial representative cross-sectional sample of suspended sediment in the river. Immediately afterward, a second set of EDIV samples were collected along with a point and a pump sample. The EDIV samples were treated as individuals and were not composited. Five additional EDIV, point, and pump samples were collected at fixed-time intervals (about 20 to 30 minutes apart). Immediately after the five additional samples were collected, a seventh set of EDIV samples were collected and later composited to provide a final representative cross-sectional sample. At the conclusion of all the EDIV, point, and pump sampling, a series of point integrated samples were collected at 20, 40, 60, and 80 percent of depth at two points in the cross-section; one within and one outside the centroid of flow (Figure 4.3.2-1).

Figure 4.3.2-1. Stream Cross-Sectional Schematic of the U.S. Geological Survey's Suspended-Sediment Sampling Scheme for the Evaluation of Samplers and Sampling Design and to Investigate Spatial and Temporal Variability (D sites are for depth-integrated verticals, B site is for the pump sample, P site is for the point sampler, and V sites are for vertical point samples at 20, 40, 60, and 80 percent of depth)

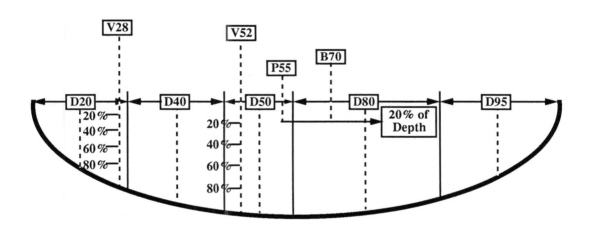

4.0 SAMPLING SEDIMENTS

4.3 Suspended sediments

4.3.2 Cross-sectional spatial and temporal variations in suspended sediment and associated trace elements and their causes

4.3.2.1 DIFFERENCES DUE TO SAMPLER TYPE OR SAMPLING DESIGN

Fluvial suspended-sediment samples are collected with different types of sampling devices and using different sampling designs. In some instances, these differences are the result of specific study/program goals; in others the differences are the result of standard organizational sampling guidelines, which are predicated on certain assumptions about the distribution of suspended sediment in fluvial cross sections. Figures 4.3.2.1-1 and 4.3.2.1-2 provide comparisons of suspended sediment and associated trace element concentrations (for the <63-μm fraction) determined on material collected with different samplers and representing different sampling designs.

In samples from the Arkansas River (Fig. 4.3.2.1-1), where the >63-μm fraction represented a substantial percentage of the suspended sediment, significant differences exist for material collected with various types of samplers/sampling designs. For example, there are significant differences in the suspended-sediment concentration, the concentration of the >63-μm fraction, and the percentages of the >63-μm and <63-μm fractions. The point-sample method seems to underestimate and the pump-sample method seems to overestimate the suspended-sediment concentration relative to the depth- and width-integrated method. These differences seem to be due to how each sampler collects the >63-μm fraction and/or where in the vertical the point and pump samplers were placed (20% of depth). In addition, there are differences between the point and pump samples with respect to the nearest EDIV sample. Again, the differences appear to result from variations in the collection efficiency of the >63-μm fraction and/or the depth at which the point and pump samplers were placed. These types of differences also appeared in other samples collected in other rivers where there was a substantial amount of >63 μm material

Although the results of the different methods showed similar concentrations of the <63-μm fractions, there were significant differences in the chemistries of this fraction. Examination of the Cu, Cr, and Ni data from the Arkansas River sample illustrates this for a pump (B-65) and a nearby EDIV (D-50) sample relative to the depth- and width-integrated sample. Similar differences for the same or other trace elements also occurred for point samples in other rivers (e.g. Fig. 4.3.2.1-2). Thus, the point or pump samples, which have been used by some investigators to represent entire cross-sectional distributions, can over- or underestimate fluvial chemical as well as suspended-sediment concentrations relative to depth- and width-integrated samples.

Differences in suspended-sediment concentration can occur even when the sample had little or no >63-μm material. Neither the point- nor pump-sample suspended-sediment concentrations are similar to those determined from the nearest EDIV sample for the West Fork Blue River (Fig. 4.3.2.1-2), despite the similarity between the point, pump, and depth- and width-integrated samples and despite the widely held view that <63-μm suspended sediment is vertically and horizontally homogeneous in fluvial cross sections (e.g., Ongley and Blachford, 1982). Also note that the point samples had elevated Cu and Zn levels relative to the depth- and width-integrated and the pump samples (Fig. 4.3.2.1-2).

The types of suspended sediment and chemical concentration differences noted above raise questions regarding the best way to obtain a representative suspended-sediment sample for subsequent physical and chemical analysis. Based on the examples provided, as well as other published data (Horowitz, et al., 1989c, Horowitz, et al., 1990), a conservative approach would seem to suggest the need for using a depth- and width-integrated isokinetic sampling technique to insure the collection of a representative sample.

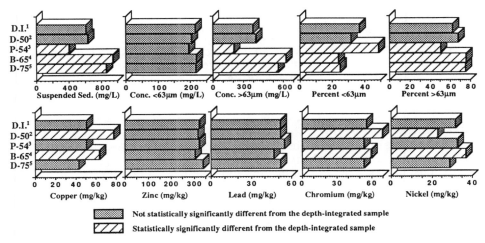

Figure 4.3.2.1-1 Comparison of the Effects of Sampling Methods on Suspended Sediment and Associated Trace Element (for the <63 μm Fraction) Concentrations for the Arkansas River, Colorado, on 5/11/87 (Data from Horowitz, et al., 1990)

Not statistically significantly different from the depth-integrated sample

Statistically significantly different from the depth-integrated sample

[1] Depth- and width-integrated isokinetic sample
[2] Equal discharge increment vertical taken 50 feet (15.25 m) from the left bank
[3] Point sample obtained at 20 percent of depth, 54 feet (16.47 m) from the left bank
[4] Pump sample obtained at 20 percent of depth, 65 feet (19.83 m) from the left bank
[5] Equal discharge increment vertical taken 75 feet (22.88 m) from the left bank

Figure 4.3.2.1-2 Comparison of the Effects of Sampling Methods on Suspended Sediment and Associated Trace Element (for the <63 μm Fraction) Concentrations for the West Fork Blue River, Nebraska on 5/27/87 (Data from Horowitz, et al., 1990)

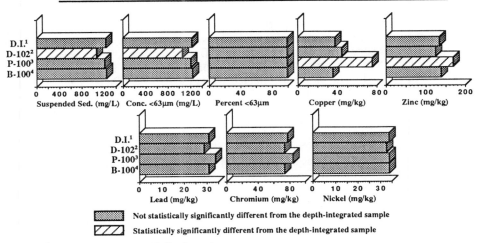

Not statistically significantly different from the depth-integrated sample

Statistically significantly different from the depth-integrated sample

[1] Depth- and width-integrated isokinetic sample
[2] Equal discharge increment vertical taken 102 feet (31.11 m) from the left bank
[3] Point sample obtained at 20 percent of depth, 100 feet (30.50 m) from the left bank
[4] Pump sample obtained at 20 percent of depth, 100 feet (30.50 m) from the left bank

4.0 SAMPLING SEDIMENTS

4.3 Suspended sediments

4.3.2 Cross-sectional spatial and temporal variations in suspended sediment and associated trace elements and their causes

4.3.2.2 HORIZONTAL VARIATIONS

All the samples at all the sites, with the exception of one (see Section 4.3.2.6), were obtained during periods of constant stage/discharge (within measurement error). Even under steady-state conditions, substantial short-term horizontal spatial variations in suspended-sediment and associated trace element concentrations were observed (Figure 4.3.2.2-1).

Figure 4.3.2.2-1 contains three sets of graphs which provide three 'snapshots' of the cross-sectional distributions of suspended sediment and associated trace elements for the Arkansas River on 5/11/87. The bars on each graph represent individual EDIV samples. The three sets of samples were obtained 20 minutes apart. First, examine the upper 5 graphs in each set. The data indicate substantial spatial differences in suspended sediment concentrations, the concentration of the >63-μm fractions, and the percentages of the >63- and <63-μm fractions. Suspended-sediment concentration shows a significant increase from the banks toward the center of the river. This difference is caused by variations in the concentration of the >63-μm material and is almost certainly a function of stream velocity which is higher toward the centroid of flow where frictional resistance to water movement is lower than near the banks. The lower velocities reduce a river's capacity for suspending and transporting larger-sized sediment particles. These velocity differences do not affect the concentrations of the <63-μm material which was essentially constant across the entire river. Similar suspended sediment distributions have been noted frequently in many other rivers (e.g., Feltz and Culbertson, 1972; Vanoni, 1977; OWDC, 1982).

Now examine the lower 3 graphs in each set (Figure 4.3.2.2-1). There were significant differences in the spatial distributions of Cu, Cr, and Ni. Please note that these differences *were not* caused by variations in the percentages of <63- and >63-μm material because the graphed concentrations are only for the <63-μm fractions. For example, the Cu concentrations for samples collected 20 (6.1 m) and 50 (15.25 m) feet from the left bank were elevated relative to the other samples in the cross section. These elevated concentrations extended into the second set of samples taken 20 minutes later. The elevated concentration of Cu 50 (15.25 m) feet from the left bank continued into the third sample set taken 40 minutes after the initial samples were collected while the elevated Cu concentration in the sample collected 20 (6.1 m) feet from the left bank did not. A similar pattern occurred in the same samples for Ni. The chemical variations noted above did not occur in subsequent samples (not shown). Please bear in mind that the physical separations between the two high Cu and Ni samples were only 30 (9.15 m) feet and that they were separated by a non-elevated sample taken 40 (12.2 m) feet from the left bank. The non-elevated sample was separated from the elevated samples by 20 (6.1 m) feet in the first, and 10 (3.05 m) feet in the second case. The Cr pattern differed from both the Cu and Ni patterns; it was elevated only in the third set of samples which was taken 40 minutes after the initial samples and was the only elevated Cr level for this entire data set. Finally, the patterns for Fe, Al, Mn, Ti, Zn and Pb (not shown) displayed no elevated concentrations for this entire data set. Bear in mind that the chemical variations in the <63-μm material occurred even though the concentration of <63-μm material was relatively constant in the cross section.

Similar spatial patterns in the concentration and grain size distributions of suspended sediments were observed throughout the study. Also, similar types of chemical variations occurred throughout the study for the same, and different trace elements. Finally, although this discussion centered on the <63-μm fraction chemical data, similar trace element spatial variations also were observed, at a similar frequency, for the >63-μm material.

Figure 4.3.2.2-1. Horizontal Variability in Suspended-Sediment and Associated Trace Element (for the <63 μm Fraction) Concentrations for the Arkansas River, Colorado, on 5/11/87 (Data from Horowitz, et al., 1990)

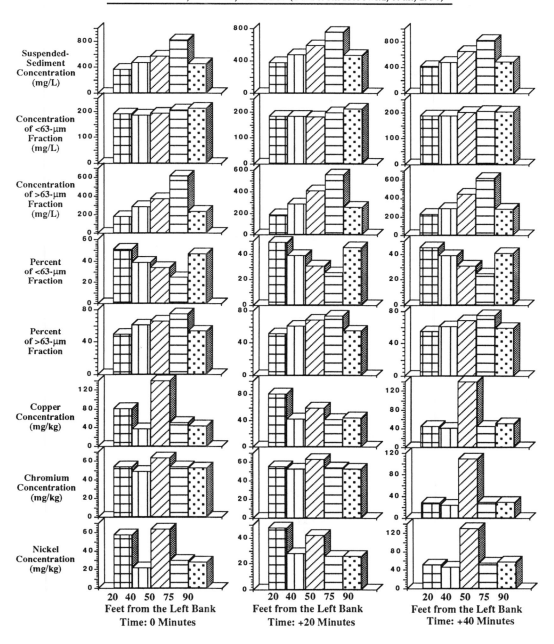

109

4.0 SAMPLING SEDIMENTS

4.3 Suspended sediments

4.3.2 Cross-sectional spatial and temporal variations in suspended sediment and associated trace elements and their causes

4.3.2.3 VERTICAL VARIATIONS

Vertical spatial variations in suspended-sediment concentration (Fig. 4.3.2.3-1) can be as marked as the horizontal spatial variations already noted in the previous section (Fig. 4.3.2.2-1). Figure 4.3.2.3-1 contains two sets of graphs which provide data on the vertical distribution of suspended sediment and associated trace elements for the Cowlitz River on 4/20/87. The bars on each graph represent individual integrated point samples taken at 20, 40, 60, and 80 percent of depth; the first set of graphs are for samples collected 195 (59.48 m) feet from the left bank while the second set of graphs are for samples collected 75 (22.88 m) feet from the left bank.

First, examine the upper 5 graphs in each set. The data indicate substantial vertical spatial differences in suspended-sediment concentration with changes in depth; as the depth increases, the concentration of the >63-μm material also increased. This is a function of the lifting capacity of the water; coarser sediment grains usually can not be raised off the stream bed, or maintained at as high an elevation off the stream bed, as fine particles. This type of observation has been reported previously for both fluvial studies in the natural environment and in controlled studies carried out in flumes (e.g., Vanoni, 1977). The vertical distribution of the <63-μm material is much more constant than that for the >63-μm material, but not as constant as found for the horizontal spatial distributions of this same-sized material described in the previous section (4.3.2.2). Suspended sediment concentrations for the samples collected at 20 and 80 percent of depth in the vertical collected 75 (22.88 m) feet from the left bank, and at 40 percent of depth in the vertical collected 195 (59.48 m) feet from the left bank are significantly different from the other samples collected at these respective locations. This reinforces the conclusion from the West Fork Blue River data (Section 4.3.2.1) which indicated that <63-μm material may not be distributed homogeneously in vertical fluvial cross sections. Despite the inconsistent patterns for the concentration of <63-μm material with increasing depth, the percentage of <63-μm material consistently decreased and the percentage of >63-μm material consistently increased with depth.

Now examine the lower 3 graphs in each set (Fig. 4.3.2.3-1). There were significant differences in the vertical distributions of Cu, Cr, and Ni. Please note that these differences *were not* caused by variations in the percentages of <63- and >63-μm material because the graphed concentrations are only for the <63-μm fractions. Cu is elevated only in the sample taken at 60 percent of depth, 75 (22.88 m) feet from the left bank. On the other hand, Zn is elevated only in the sample taken at 60 percent of depth, 195 (59.48 m) feet from the left bank. Finally, Pb is depressed in the 60 percent of depth sample, and elevated in the 80% depth of sample, 195 (59.48 m) feet from the left bank. Similar chemical patterns with depth also occur for these and other trace elements in different rivers, and for the >63-μm fractions in this and other rivers.

Do not assume that the significant differences in suspended sediment and associated trace element concentrations described for the Cowlitz River occurred because of substantial differences in the vertical separation between the samples. River depth at the 195 foot (59.48m) vertical was only 8.1 (2.47m) feet and 8.6 (2.62m) feet at the 75 foot (22.88m) vertical. Thus, the physical separation between the point samples in each vertical averaged less than 2 (0.61m) feet (Fig. 4.3.2.3-1).

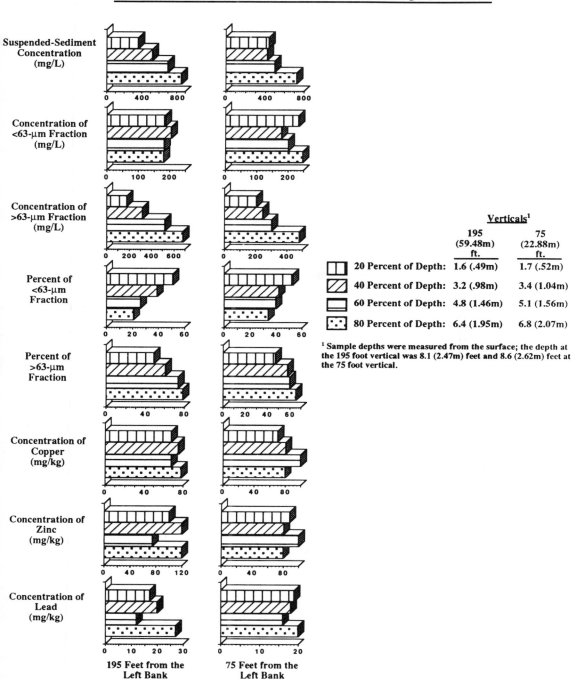

Figure 4.3.2.3-1. Vertical Spatial Variability in Suspended-Sediment and Associated Trace Element (for the <63 μm Fraction) Concentrations from Integrated Point Samples Taken at Selected Depths in the Cowlitz River, Washington, on 4/20/87

Verticals[1]

		195 (59.48m) ft.	75 (22.88m) ft.
☐	20 Percent of Depth:	1.6 (.49m)	1.7 (.52m)
▨	40 Percent of Depth:	3.2 (.98m)	3.4 (1.04m)
☐	60 Percent of Depth:	4.8 (1.46m)	5.1 (1.56m)
⬚	80 Percent of Depth:	6.4 (1.95m)	6.8 (2.07m)

[1] Sample depths were measured from the surface; the depth at the 195 foot vertical was 8.1 (2.47m) feet and 8.6 (2.62m) feet at the 75 foot vertical.

4.0 SAMPLING SEDIMENTS

4.3 Suspended sediments

4.3.2 Cross-sectional spatial and temporal variations in suspended sediment and associated trace elements and their causes

4.3.2.4 IMPORTANCE OF SILT/CLAY VERSUS SAND-SIZED SUSPENDED SEDIMENT FOR TRACE ELEMENT TRANSPORT

As already stated, grain size exercises a disproportionate control on sediment-associated trace element concentrations; as grain size decreases, trace element concentrations increase (see Sections 2.2.1 to 2.2.6). Also, as noted in Section 2.2.6, as discharge increases, the concentration and percentage of coarser-sized material, less rich in trace elements increases. As a result, under most circumstances, as discharge increases, the overall concentration of suspended sediment-associated trace elements should decrease (see Section 2.2.6). How then do these interrelations affect sediment-associated trace element concentrations and transport ? It could be inferred that maximum trace element concentrations and transport occurs during low discharge because trace element concentrations are highest at the same time. A corollary to this inference could be that the >63-μm material could probably be ignored, if not for suspended sediment transport itself, then at least for sediment-associated trace element transport. However, the increased concentrations that occur during low flow have to be weighed against the substantial increases in sediment mass that are transported during periods of high discharge.

In samples collected at the Arkansas and Cowlitz River sites, the >63-μm material represented substantial percentages of the collected suspended sediment (Table 4.3.2.4-1). The data indicate that the >63-μm material should not be ignored in terms of its contribution to the overall trace element concentrations in the sample, even when the trace element concentrations associated with the <63-μm material are twice as high or higher than those associated with the >63-μm material (e.g., Mn, Cu, Zn, Pb, and Ni for the Arkansas River). Thus, because the >63-μm material contributes substantially to the overall sample concentration, it can not be ignored when calculating annual sediment-associated trace element transport at these sites. Unfortunately, without additional samples in which the >63-μm fraction represents a smaller percentage (e.g., less than some 60 percent) of the overall sample, it is difficult to assess at what point the >63-μm contributions become statistically insignificant. However, a series of hypothetical mixing calculations using the Arkansas River size-fraction chemical data indicate that concentrations as low as 10 to 15 percent of the >63-μm material would make substantial contributions (greater than 10 percent) to at least some of the total sediment-associated trace element concentrations.

The data from the Cowlitz River is interesting because it appears as if there is only a limited grain-size affect on trace element concentrations (Table 4.3.2.4-1). Only Cu, Cr, and Ni appear to be affected by grain size whereas in the Arkansas River samples, all the trace elements except Al displayed the effect. Even those Cowlitz trace elements that do appear to display the grain size effect, do not demonstrate it with the same magnitude as the Arkansas trace elements. Although there are probably several causes for this result, the major one is that there is only a single dominant source for suspended sediment in the Cowlitz River. Suspended sediment at this site is composed almost entirely of recent volcanic material from Mt. St. Helens. Thus, despite the differences in grain size, all the material is compositionally similar; therefore, the grain size effect is substantially minimized.

Table 4.3.2.4-1. Grain-Size Fractional Contributions for Suspended Sediment for Selected Rivers (Data from Horowitz, et al., 1990)

Arkansas River, Colorado 5/11/87[*]
(<63-μm fraction = 37%, >63-μm fraction = 63%)

Element	Trace Element Concentrations in			Percent Contributions of	
	<63-μm Fraction	>63-μm Fraction	Total Sample	<63-μm Fraction	>63-μm Fraction
Fe (%)	4.7	3.5	3.9	44	56
Al (%)	7.1	7.6	7.4	35	65
Ti (%)	0.41	0.33	0.36	42	58
Mn (mg/kg)	1100	600	800	50	50
Cu (mg/kg)	51	22	33	58	42
Zn (mg/kg)	325	110	190	63	37
Pb (mg/kg)	52	25	35	54	46
Cr (mg/kg)	56	44	49	43	57
Ni (mg/kg)	32	16	22	55	45
Co (mg/kg)	15	11	12.5	45	55

Cowlitz River, Washington 4/20/87[*]
(<63-μm fraction = 41%, >63-μm fraction = 59%)

Element	Trace Element Concentrations in			Percent Contributions of	
	<63-μm Fraction	>63-μm Fraction	Total Sample	<63-μm Fraction	>63-μm Fraction
Fe (%)	3.8	4.0	4.0	40	60
Al (%)	8.3	8.8	8.6	40	60
Ti (%)	0.44	0.41	0.41	40	60
Mn (mg/kg)	650	670	660	40	60
Cu (mg/kg)	63	33	46	57	43
Zn (mg/kg)	62	68	59	42	58
Pb (mg/kg)	12	10	10.8	45	55
Cr (mg/kg)	35	19	25	56	44
Ni (mg/kg)	25	16	19	53	47
Co (mg/kg)	14	14	14	41	59

[*] The data in these tables represent the mean of the initial and final composite samples obtained at these sampling sites.

113

4.0 SAMPLING SEDIMENTS

4.3 Suspended sediments

4.3.2 Cross-sectional spatial and temporal variations in suspended sediment and associated trace elements and their causes

4.3.2.5 SUSPENDED SEDIMENT AND ASSOCIATED TRACE ELEMENT TEMPORAL VARIATIONS DURING CONSTANT DISCHARGE

Short-term temporal variability during constant stage/discharge also was observed in a number of samples (Table 4.3.2.5-1 and Fig. 4.3.2.5-1). For example, significant differences in suspended-sediment concentrations exist for the first composite and the first five calculated composite samples relative to the sixth calculated and the second composite samples (a maximum increase of nearly 60%) for the 5/11/87 Arkansas River sediments (Table 4.3.2.5-1). There are no significant differences in the percentages of the >63-μm and <63-μm material, nor in the chemistry of the samples; however, significant differences in the concentrations of both the >63-μm and <63-μm material did occur, and produced a significant increase in suspended-sediment concentration. Thus, a significant increase in suspended-sediment concentration occurred during a period of constant stage/discharge.

The data from the 5/11/87 Arkansas River and the West Fork Blue River samples (Section 4.3.2.1) raise an important question about suspended sediment distributions and dynamics. It is generally held that the transport and cross-sectional distribution of coarse (>63μm, sand-sized) suspended sediment is heterogeneous and controlled by discharge while the transport and cross-sectional distribution of fines (<63μm, silt/clay) is homogeneous and transport is controlled by supply (e.g., Vanoni, 1977). The Arkansas River data show that a significant increase in both coarse and fine suspended sediment occurred when no significant increase in discharge occurred. The West Fork Blue River data indicate that fine material may not be homogeneously distributed, at least vertically, in a cross section. Therefore, the view that coarse suspended sediment concentrations, distributions, and transport are controlled solely by discharge, while fine suspended sediment concentrations, distributions, and transport are controlled solely by supply probably should be viewed as an oversimplification.

Examination of individual EDIV sample data for the 5/11/87 Arkansas River sediments indicates that there was short-term temporal variability in suspended-sediment concentration, in the concentration of the >63-μm and the <63-μm fractions, and in Cu, Cr, and Ni concentrations (Figure 4.3.2.5-1). For example, Cu concentrations for the first two samples obtained 20 feet (6.1m) from the left bank were elevated for the first 40 minutes of sampling relative to the next four (from 40 to 120 minutes into sampling). The elevated Cu concentrations noted at this site were not detected in simultaneously collected samples 40 feet (12.2m) from the left bank, only 20 feet (6.1m) away; however, they were detected at the site 50 feet (15.25m) from the left bank, which is 30 feet (9.15m) from the 20 foot site and 10 feet (3.05m) from the 40 foot site. It seems that <63μm sediment-associated trace elements can be distributed heterogeneously despite the fact that the <63μm sediment itself is homogeneously distributed in fluvial cross sections . Similar heterogeneous distributions occur for Cr, and Ni in the Arkansas River samples, and in additional samples and for other trace elements collected from other sites during this study, (Horowitz, et al., 1990). Finally, although the chemical data in Figure 4.3.2.5-1 were for the <63-μm fractions, and the discussion centered on these materials, similar temporal trace element variations also were observed at a similar frequency for the >63-μm material.

Table 4.3.2.5-1. Short Term Temporal Variability in Suspended Sediment and Associated Trace Elements (<63-μm Fraction) Concentrations During Constant Discharge[1] for the Arkansas River, Colorado, on 5/11/87 (Data from Horowitz, et al., 1990)

Sample	S. Sed. Concen. (mg/L)	Concentration of S. Sed (mg/L) <63 μm	>63μm	Percentage of <63 μm	>63μm	Fe (%)	Al (%)	Cu (mg/kg)	Zn (mg/kg)	Pb (mg/kg)	Cr (mg/kg)	Ni (mg/kg)
Com-1	566	204	362	36	64	4.5	6.8	42	320	50	49	24
C.Com-1	539	197	341	39	61	4.7	7.1	71	330	51	55	40
C.Com-2	534	192	342	38	62	4.8	7.3	54	349	55	56	34
C.Com-3	575	198	377	36	64	4.7	7.3	65	317	51	68	44
C.Com-4	545	194	351	38	62	4.6	7.1	44	323	52	54	34
C.Com-5	577	196	382	37	63	4.7	7.4	45	303	52	53	26
C.Com-6	725[2]	267	458	39	61	4.6	7.2	49	278	46	56	28
Com-1	886	284	602	32	68	4.9	6.7	39	380	61	59	28

[1]Discharge during the sampling operation remained constant at 1995 ft³/s (56.5m³/s) with no change in stage 4.27 ft (1.30 m)
[2]Underlined data are significantly different from the others in the same column

Figure 4.3.2.5-1. Short-Term Temporal Variations in Suspended Sediment and Associated Trace Element (<63-μm Fraction) Concentrations During Constant Discharge for the Arkansas, River, Colorado, for 5/11/87 (Data from Horowitz, et al., 1990)

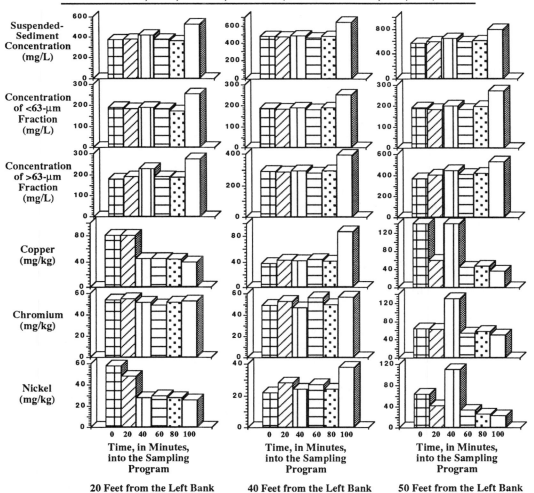

Suspended-Sediment Concentration (mg/L)		
Concentration of <63-μm Fraction (mg/L)		
Concentration of >63-μm Fraction (mg/L)		
Copper (mg/kg)		
Chromium (mg/kg)		
Nickel (mg/kg)		

Time, in Minutes, into the Sampling Program

20 Feet from the Left Bank   40 Feet from the Left Bank   50 Feet from the Left Bank

115

4.0 SAMPLING SEDIMENTS

4.3 Suspended sediments

4.3.2 Cross-sectional spatial and temporal variations in suspended sediment and associated trace elements and their causes

4.3.2.6 SUSPENDED SEDIMENT AND ASSOCIATED TRACE ELEMENT TEMPORAL VARIATIONS DURING CHANGES IN DISCHARGE

While sampling the Arkansas River on 5/29/87, a thunderstorm occurred upstream from the sampling site. Runoff from the storm increased the turbidity in the river during and after collection of the fourth set of samples (Table 4.3.2.6-1). The effects of the thunderstorm are reflected in the analyses for the fifth (C.Com-5) and particularly the sixth (C.Com-6) set of simultaneous EDIV samples. In the fifth set, the most apparent effects are increased concentrations of Pb and Zn, and to a lesser extent Cu. In the sixth set, the effects displayed for the same elements are more substantial. Comparison of data from the fifth and sixth sets indicates that suspended-sediment concentrations increased 26%, the concentration of <63-μm material increased 60%, and the concentrations of Cu, Zn, and Pb increased 2- to 10-fold. Surprisingly, although the percentage changed, the concentration of the >63-μm material remained either constant or decreased slightly. Thus, it would seem that the observed changes in suspended-sediment chemistry were more the result of a significant increase in the percentage of fine-grained sediment than to a change in total suspended sediment concentration (although this did increase). Please note that increases in suspended-sediment concentration are typical when there is an increase in stream velocity/discharge (e.g., Vanoni, 1977; OWDC, 1982); however, the usual cause is an increase in the concentration of the >63-μm (sand-sized) material rather than an increase in the <63-μm material (silt/clay-sized). Throughout the sampling operation at this site, the concentrations of the major elements (Fe, Mn, Al, and Ti) and some of the trace elements (Cr, Ni, and Co) for the <63-μm material remained constant (within analytical error). The increases noted in the sixth set of EDIV samples also continue into the final cross-sectional composite sample (Com-2). It should be noted that the time between the collection of the fourth and sixth simultaneous EDIV samples was about 40 to 60 minutes; the final composite sample (Com-2) was obtained immediately after the sixth set of EDIV samples were collected .

Under normal sampling conditions, a depth- and width-integrated sample is obtained by a one- or two-man crew, and sample collection takes between 0.5 and 2 hours. If a sample had been collected at the time of the first composite or 2 to 3 hours after the final composite (Com-2), the thunderstorm-effected changes to suspended-sediment concentration and chemistry probably would have been missed entirely. If sampling began around the time that the third or fourth EDIV samples were collected, then only a portion of the storm's contribution would have been collected and quantitated. These data highlight the potential effects of an 'event' on suspended sediment and associated trace element concentrations. The changes obviously were important in terms of daily discharge and transport. Their importance on an annual basis also could be significant or insignificant, depending on local conditions, and would have to be evaluated on the basis of an annual record. On the other hand, significance could not be evaluated if no sample at all had been obtained.

Table 4.3.2.6-1. Suspended Sediment and Associated Trace Element Variations During Changes in Discharge: Arkansas River, Colorado, 6/29/87 (Data from Horowitz, et al., 1990)

		Concentrations of		Percent of		Trace Element Concentrations in the <63-μm Fractions			
		<63-μm	>63-μm						
	S. Sed.	Fraction	Fraction	<63-μm	>63-μm	Fe	Cu	Zn	Pb
Sample	(mg/L)	(mg/L)	(mg/L)	Fraction	Fraction	(Wt. %)	(mg/kg)	(mg/kg)	(mg/kg)
Com-1[a]	N/A[c]	108	N/A	N/A	N/A	3.6	41	340	54
C. Com-1[b]	322	105	217	37	63	3.6	42	436	71
C. Com-2[b]	356	126	229	41	59	3.4	39	428	67
C. Com-3[b]	346	146	200	48	52	3.5	39	374	57
C. Com-4[b]	321	131	190	49	51	3.4	40	394	63
C. Com-5[b]	306	131	175	48	52	3.6	47	468	<u>118</u>
C. Com-6[b]	<u>387</u>[d]	<u>210</u>	177	<u>60</u>	<u>40</u>	3.9	<u>75</u>	<u>923</u>	$\geq$<u>1300</u>
Com-2[a]	<u>406</u>	<u>245</u>	161	<u>60</u>	<u>40</u>	3.9	<u>74</u>	<u>$\geq$1000</u>	<u>$\geq$1300</u>

[a] - actual composites obtained by combining EDIV samples

[b] - calculated composites obtained by mathematically combining the results from EDIV samples

[c] - not available

[d] - underlined numbers are significantly different from the others in the same column

5.0 Summary and General Considerations

Having reached this point in the Primer, you have been exposed to most of the basic principles covering the complex subject of sediment-trace element chemistry. We have discussed why sediment chemistry should be an integral part of most water-quality studies. We also have mentioned the major physical and chemical (geochemical) factors which control sediment-trace element chemistry, and discussed in some detail how these various factors can be determined. By now, it also should be apparent that, although the geochemical factors were covered as separate entities, they are all strongly interrelated. Finally, we have discussed the problems and considerations involved in collecting both bottom and suspended sediment for subsequent chemical analysis.

The intent of this Primer is to introduce you to the highly complex subject of sediment-trace element chemistry; not to make you an expert in the field. If one or several of the covered subjects is of particular interest, follow up the introductory material in the text by consulting the appropriate references listed in the following section. Many of them were consulted and used in developing the contents of this Primer, but many are listed because they more completely describe methods and techniques or because they provide some excellent examples of how sediment chemistry can be used to deal with various types of environmental studies. Several also provide excellent examples of how to conduct various types of sediment-chemical investigations.

Use and build upon the work of others. Although no one receives credit for 'reinventing the wheel', there is nothing wrong with using a sampling and interpretational approach that successfully has been applied elsewhere. You might save some time, initially, by ignoring previous work, during the planning of a study, or during data analysis and report preparation. However, most studies do not produce major new advances. Remember, the purposes of good science, and of most studies, are to answer specific questions as capably as possible within time and financial constraints—not to find the ultimate solution.

I have spent the last twenty years of my professional life working in the field of sediment-trace element chemistry and I honestly believe that we are no closer to a complete understanding of this complex subject today than we were twenty years ago. It seems that the more we find out, the more we discover what we do not know, and need to know. The more studies we carry out to answer specific questions, the more new questions we raise. To some degree this can be highly discouraging; however, that is the nature of this, and many other scientific disciplines; it comes 'with the territory', and makes sediment-chemical studies perpetually interesting and challenging.

Selected References

*   Ackermann, F., Bergmann, H., and Schleichert, U., 1983, Monitoring of heavy metals in coastal and estuarine sediments - a question of grain size: <20 μm versus <60 μm: Environmental Technology Letters, v. 4, p. 317-328.
*   Ahlf, W., Calmano, W., Erhard, J., and Forstner, U., 1989, Comparison of five bioassay techniques for assessing sediment-bound contaminants: Hydrobiologia, v. 188/189, p. 285-289.
    Armstrong, F., 1958, Inorganic suspended matter in seawater: Journal of Marine Research, v. 17, p. 23-25.
*   Arrhenius, G., and Korkish, J., 1959, Uranium and thorium in marine minerals: First International Oceanographic Congress, American Association for the Advancement of Science, preprints, 497 p.
    Ashley, H., 1979, Particle size analysis by elutriation and centrifugation as exemplified by the Bahco Microparticle Classifier, in Stockham, J., and Fochtman, E., eds., Particle size analysis: Ann Arbor, Mich., Ann Arbor Press, p. 101-110.
*   ASTM, 1990, Standard guide for sampling fluvial sediment in motion, in, Annual book of ASTM standards, Philadelphia, PA, American Society for Testing and Materials, v. 11.02, p. 595-611.
*   Aston, S., Bruty, D., Chester, R., and Padgham, R., 1973, Mercury in lake sediments: a possible indicator of technological growth: Nature (London), v. 241, p. 450-451.
    Aston, S., Thornton, I., Webb, J., Purves, J., amd Milford, B., 1974, Stream sediment composition: an aid to water quality assessment, Water, Air, and Soil Pollution, v. 3, p. 321-325.
    Aulio, K., 1980, Accumulation of copper in fluvial sediments and yellow water lillies (*Nuphar lutea*) at varying distances from a metal processing plant: Bulletin of Environmental Contamination and Toxicology, v. 25, p. 713-717.
*   Baccini, P., Greider, E., Stierli, R., and Goldberg, S., 1982, The influence of natural organic matter on the adsorption properties of mineral particles in lake water: Schweiz Zeitschrift fuer Hydrologie, v. 44, p. 99-116.
*   Baker, R., 1980, Contaminants and sediments, v. 1 and 2, 558 p., 627 p.
*   Baldi, F., and Bargagli, R., 1982, Chemical leaching and specific surface area measurements of marine sediments in the evaluation of mercury contamination near cinnabar deposits: Marine Environmental Research, v. 6, p. 69-82.
    Balls, P.W., 1986, Composition of suspended particulate matter from Scottish coastal waters - geochemical implications for the transport of trace metal contaminants: The Science of the Total Environment, v. 57, p. 171-180.
*   Banat, K., Forstner, U., and Muller, G., 1972, Schwermetalle in sedimenten von Donau, Rhein, Ems, Weser, und Elbe im bereich der Bundesrepublik Deutschland: Naturwissenschaften, v. 12, p. 525-528.
    Barbanti, A., and Sighinolfi, G., 1988, Sequential Extraction of phosphorus and heavy metals from sediments: methodological considerations: Environmental Technology Letters, v. 9, p. 127-134.
*   Beeson, R., 1984, The use of fine fractions of stream sediments in geochemical exploration in arid and semi-arid terrains, Journal of Geochemical Exploration, v. 22, p. 119-132.
    Benjamin, M., Hayes, K., and Leckie, J., 1982, Removal of toxic metals from power-generation waste streams by adsorption and coprecipitation: Journal of the Water Pollution Control Federation, v. 54, p. 1472-1481.

---

* References cited in the text, tables, or diagrams are marked with an asterisk (*).

* Benjamin, M., and Leckie, J., 1981, Conceptual model for metal-ligand-surface interactions during adsorption: Environmental Science and Technology, v. 15, p. 1050-1057.

Benjamin, M., and Leckie, J., 1982, Effects of complexation by Cl, SO$_4$, and S$_2$O$_3$ on adsorption behavior of Cd on oxide surfaces: Environmental Science and Technology, v. 16, p. 162-170.

Berner, R., 1981, A new geochemical classification of sedimentary environments: Journal of Sedimentary Petrology, v. 51, p.359-365.

Boniforti, R., Bacciola, D., Niccolai, I., and Ruggiero, R., 1988, Selective extraction as an estimate of bioavailability of As, Cd, Co, Cr, Cu, Fe, Mn, Ni, Pb, and Zn in marine sediments collected from the central Adriatic Sea: Environmental Technology Letters, v. 9, p. 117-126.

Bostrom, K., Burman, J., Ponter, C., and Ingri, J., 1981, Selective removal of trace elements from the Baltic by suspended matter: Marine Chemistry, v. 10, p. 335-354.

* Brannon, J., Engerl, R., Rose, J., Hunt, P., and Smith, I., 1976, Distribution of Mn, Ni, Zn, Cd, and As in sediments and the standard elutriate: U.S. Army Engineers Waterways Experimental Station, Environmental Effects Laboratory, Miscellaneous Papers, D 76-18, 38 p.

Breward, N., and Peachy, D., 1983, The development of a rapid scheme for the elucidation of the chemical speciation of elements in sediments: The Science of the Total Environment, v. 29, p. 155-162.

* Bruland, K., Bertine, K., Koide, M., and Goldberg, E., 1974, History of metal pollution in the southern California coastal zone: Environmental Science and Technology, v. 8, p. 425-432.

* Brunauer, S., Emmett, P., and Teller, E., 1938, Adsorption of gases in multimolecular layers: Journal of the American Chemical Society, v. 60, p. 309-319.

*Bunzl, K., Schmidt, W., and Sanson, B., 1976, Kinetics of ion exchange in soil organic matter IV, adsorption and desorption of Pb^{+2}, Cu^{+2}, Cd^{+2}, Zn^2, and Ca^{+2}, by peat: Journal of Soil Science, v. 27, p. 32-41.

Burton, J., 1978, The modes of association of trace metals with certain compounds in the sedimentary cycle, in Goldberg, E., ed., Biogeochemistry of estuarine sediments, Paris, UNESCO Press, p. 33-41.

* Buser, W., and Graf, F., 1955, Differenzierung von mangan (II)-manganit and δ-MnO$_2$ durch oberflachenmessung nach Brunauer-Emmet-Teller: Helvetica Chimica Acta, v. 38, p. 830-842.

* Calmano, W., Ahlf, A., and Forstner, U., 1988, Study of metal sorption/desorption processes on competing sediment components with a multichamber device: Environmental Geology and Water Sciences, v. 11, p. 77-84.

Calmano, W., and Forstner, U., 1983, Chemical extraction of heavy metals in polluted river sediments in central Europe: The Science of the Total Environment, v., 28, p. 77-90.

Calmano, W., Wellershaus, S., and Forstner, U., 1982, Dredging of contaminated sediments in the Weser estuary: chemical forms of some heavy metals: Environmental Technology Lettters, v. 3, p. 199-208.

* Cameron, E., 1974, Geochemical methods of exploration for massive sulphide mineralization in the Canadian shield, in, Elliott, I., and Fletcher, W., eds., Geochemical exploration—Proceedings of the 5th international geochemical exploration symposium, Vancouver, p. 21-49.

* Campbell, P., and Tessier, A., 1989, Biological availability of metals in sediments: analytical approaches, in Vernet, J-P, ed., Heavy Metals in the Environment, v. 1, Geneva, 1989, CEP Consultants Ltd., Edinburgh, p. 516-525.

* Cantillo, A., 1986, Standard and reference materials for use in marine science, Rockville, MD, Department of Commerce, National Oceanic and Atmospheric Administration, 157 p.

* Carpenter, J., Bradford, W., and Grant, V., 1975, Processes affecting the composition of estuarine waters (HCO_3, Fe, Mn, Zn, Cu, Ni, Cr, Co, and Cd): Estuarine Research, v. 1, 188-214.

Chao, T., 1972, Selective dissolution of manganese oxides from soils and sediments with acidified hydroxylamine hydrochloride: Soil Science Society of America Proceedings, v. 36, p. 764-768.

* Chao, T., 1984, Use of partial dissolution techniques in geochemical exploration: Journal of Geochemical Exploration, v. 20, 101-136.

* Chao, T., and Zhou, L., 1983, Extraction techniques for selective dissolution of amorphous iron oxides from soils and sediments: Soil Science Society of America Journal, v. 47, p. 225-232.

* Chapman, P., Romberg, G., and Vigers, G., 1982, Design of monitoring studies for priority pollutants: Journal of the Water Pollution Control Federation, v. 54, p. 292-297.

Charm, W., 1967, Freeze drying as a rapid method of disaggregating silts and clays for particle size analysis: Journal of Sedimentary Petrology, v. 37, p. 970-971.

* Chen, K., Gupta, S., Sycip, A., Lu, J., Knezevic, M., and Choi, W., 1976, The effect of dispersion and resedimentation on migration of chemical constituents during open water disposal of dredged material: Contract Report, U.S. Army Engineers Waterways Experimental Station, Vicksburg, Miss., 221 p.

* Chester, R., 1965, Geochemical criteria for differentiating reef from non-reef facies in carbonate rocks: American Association of Petroleum Geologists Bulletin, v. 49, p. 253-276.

* Chester, R., and Hughes, M., 1966, The distribution of manganese, iron, and nickel in a north Pacific deep sea clay core: Deep Sea Research, v. 13, p. 627-634.

* Chester, R., and Hughes, M., 1967, A chemical technique for the separation of ferromanganese minerals, carbonate minerals, and adsorbed trace elements from pelagic sediments: Chemical Geology, v. 2, p. 249-262.

* Chester, R., and Hughes, M., 1969, The trace element geochemistry of a north Pacific pelagic clay core: Deep Sea Research, v. 16, p. 639-654.

* Chester, R., and Messiah-Hanna, R., 1970, Trace element partition patterns in North Atlantic deep sea sediments: Geochimica et Cosmochimica Acta, v. 34, p. 1112-1128.

* Chester, R., and Stoner, J., 1975, Average trace element composition of low level marine atmospheric particulates: Nature (London), v. 246, p. 138-139.

Chester, R., Thomas, A., Lin, F., Basaham, A., and Jacinto, G., 1988, The solid state speciation of copper in surface water particulates and oceanic sediments: Marine Chemistry, v. 24, p. 261-292.

* Copeland, R., 1972, Mercury in the Lake Michigan environment, in, Hartung, R., and Dinman, B., eds., Environmental mercury contamination, Ann Arbor, Ann Arbor Science Publishers, p. 71-76.

* Cronan, D., 1974, Authigenic minerals in deep sea sediments, in Goldberg, E., ed., The Sea, v. 5, New York, John Wiley, p. 491-525.

* Cronan, D., and Garrett, D., 1973, The distribution of elements in metalliferous Pacific sediments collected by the DSDP: Nature Physical Science, v. 242, p. 88-89.

* Curlick, J., Fulajtar, E., Glinski, J., and Michalowska, K., 1974, Relation between the surface area and some properties of silty soils of Poland and Czechoslovakia: Polish Journal of Soil Science, v. VI, p. 12-19.

Davies, B., and Wixson, B., 1987, Use of factor analysis to differentiate pollutants from other trace metals in surface soils of the mineralized area of Madison County, Missouri, U.S.A.: Water, Air, and Soil Pollution, v. 33, p. 339-348.

Davies-Colley, R., Nelson, P., and Williamson, K., 1984, Copper and cadmium uptake by estuarine sedimentary phases: Environmental Science and Technology, v. 18, p. 491-499.

* Davis, C., Ahmad, N., and Jones, R., 1971, Effect of exchangeable cations on the surface area of clays: Clay Mineralogy, v. 9, p. 258-261.

* Davis, J., and Leckie, J., 1978, Effects of adsorbed complexing ligands on trace metal uptake by hydrous oxides: Environmental Science and Technology, v. 12, p. 1309-1315.

* de Groot, A., Zshuppe, K., and Salomons, W., 1982, Standardization of methods of analysis for heavy metals in sediments: Hydrobiologia, v. 92, p. 689 - 695.

Deurer, R., Forstner, U., and Schmoll, G., 1978, Selective chemical extraction of carbonate-associated metals from recent lacustrine sediments: Geochimica et Cosmochimica Acta, v. 42, p. 415-427.

* Diks, D., and Allen, H., 1983, Correlation of copper distribution in a freshwater-sediment system to bioavailability: Bulletin of Environmental Contamination and Toxicology, v. 30, p. 37-43.

* Dossis, P., and Warren, L., 1980, Distribution of heavy metals between the minerals and organic debris in a contaminated marine sediment, in Baker, R., ed., Contaminants and sediments, v. 1, Ann Arbor, Mich., Ann Arbor Science Publishers, Inc., p. 119-142.

* Drever, J., 1982, The geochemistry of natural waters: Englewood Cliffs, N.J., Prentice-Hall, Inc., 333 p.

* Duchart, P., Calvert, S., and Price, N., 1973, Distribution of trace metals in the pore waters of shallow water marine sediments: Lmnology and Oceanography, v. 18, p. 605-610.

Durum, W., and Haffty, J., 1963, Implications of minor element content of some major streams of the world: Geochimica et Cosmochimica Acta, v. 27, p. 1-11.

* Dymond, J., Corliss, J., Heath, G., Field, C., Dasch, J., and Veeh, H., 1973, Origin of metalliferous sediments from the Pacific ocean: Geological Society of America Bulletin, v. 84, p. 3355-3372.

* Egashira, K., Kajiyama, T., and Arimizu, N., 1977, Effects of sodium dithionate-sodium bicarbonate-sodium citrate and 2% sodium carbonate treatments on the surface area of allophane and imogolite: Nendo Kagaku, v. 17, p. 38-47.

* Elrick, K., and Horowitz, A., 1986, Analysis of rocks and sediments for As, Sb, and Se by wet digestion, atomic absorption spectroscopy, and hydride generation, Varian Instruments at Work, AA-56, 5 p.

* Elrick, K., and Horowitz, A., 1987, Analysis of rocks and sediments for mercury by wet digestion and flameless cold vapor atomic absorption, Varian Instruments at Work, AA-72, 5 p.

Engler, R., Brannon, J., and Rose, J., 1974, A practical selective extraction procedure for sediment characterization: American Chemical Society, v. 168, Atlantic City, N.J., 17 p.

* EPRI, 1985, Sampling design for aqautic ecologic monitoring, Palo Alto, CA, Electric Power Research Institute, v. 1-5, EPRI EA-4302.

* Etchebar, H., and Jouanneau, J., 1980, Comparison of different methods for the recovery of suspended matter from estuarine waters: deposition, filtration, and centrifugation: consequences for the determination of some heavy metals: Estuarine and Coastal Marine Science, v. 11, p. 701-707.

* Fauth, H., Hindel, R., Siewers, U., and Zinner, J., 1985, Geochimischer Atlas Bundesrepublik Deutschland - Verteilung von Schwermetallen in Wassern und Bachsedimenten, Hannover, Bundesanstatt fur Geowissenschaften und Rohstoffe, 79 p.

* Feltz, H., 1980, Significance of bottom material data in evaluating water quality, in Baker, R., ed., Contaminants and sediments, v. 1, Ann Arbor, Mich., Ann Arbor Science Publishers, Inc., p. 271-287.

* Feltz, H., and Culbertson, J., 1972, Sampling procedures and problems in determining pesticide residues in the hydrologic environment: Pesticide Monitoring Journal, v. 6, p. 171-178.

* Filipek, L., Chao, T., and Carpenter, J., 1981, Factors affecting the partitioning of Cu, Zn, and Pb in boulder coatings and stream sediments in the vicinity of a polymetallic sulfide deposit: Chemical Geology, v. 33, p. 45-64.

* Filipek, L., and Owen, R., 1979, Geochemical associations and grain size partitioning of heavy metals in lacustrine sediments: Chemical Geology, v. 26, p. 105-117.

* Flanagan, F., 1976, Description of eight new USGS rock standards, U.S. Geological Survey Professional Paper 840, Washington, D.C., U.S. Government Printing Office, 192 p.

* Folk, R., 1966, A review of grain-size parameters: Sedimentology, v. 6, p. 73-93.

* Forstner, U., 1982a, Chemical forms of metal enrichment in recent sediments, in Amstutz, G., and others, eds., Ore genesis, New York, Springer-Verlag, p. 191-199.

* Forstner, U., 1982b, Accumulative phases for heavy metals in limnic sediments: Hydrobiologia, v. 91, p. 269-284.

* Forstner, U., 1989, Contaminated sediments, in Bhattacharji, S. and others, Lecture notes in earth sciences, v. 21, New York, Springer-Verlag, 157 p.

* Forstner, U., Ahlf, W., and Calmano, W., 1989, Studies on the transfer of heavy metals between sedimentary phases with a multi-chamber device: combined effects of salinity and redox variation: Marine Chemistry, v. 28, p. 145-158.

Forstner, U., and Patchineelam, S., 1980, Chemical associations of heavy metals in polluted sediments from the lower Rhine River, in Kavanaugh, M., and Leckie, J., eds., Particulates in water, Advances in chemistry series, v. 189, American Chemical Society, Washington, D.C., p. 177-193.

* Forstner, U., and Salomons, W., 1984, Metals in the hydrocycle, New York, Springer-Verlag, 349 p.

* Forstner, U., and Stoffers, P., 1981, Chemical fractionation of transition elements in Pacific pelagic sediments: Geochimica et Cosmochimica Acta, v. 45, p. 1141-1146.

* Forstner, U., and Wittmann, G., 1981, Metal pollution in the aquatic environment, 2nd revised edition, New York, Springer-Verlag, 486 p.

* Francis, C., Bonner, W., and Tamura, T., 1972, An evaluation of zonal centrifugation as a research tool in soil science, I, methodology, Soil Science Society of America Proceedings, v. 36, p. 366-376.

Francis, C., and Brinkley, F., 1976, Preferential adsorption of Cs^{137} to micaceous minerals in contaminated freshwater sediment: Nature (London), v. 260, p. 511-513.

* Fripiat, J., and Gastuche, M., 1952, Etude physiochimique des surfaces des argiles - Les combinaisons de la kaolinite avec des oxides de fer trivalent: Publications de L'Institute National Pour L'Etude Agronomique du Congo, Belge, v. 54, p. 7-35.

* Gambrell, R., Khalid, R., Verloo, M., and Patrick, W., 1977, Transformations of heavy metals and plant nutrients in dredged sediments as affected by oxidation-reduction potential and pH: U.S. Army Corps of Engineers, Vicksburg, Miss., Report D 77-4, 309 p.

* Ghosh, K., and Schnitzer, M., 1981, Fluorescense excitation spectra and viscosity behavior of a fulvic acid and its copper and iron complexes: Soil Science Society of America Journal, v. 45, p. 25-29.

Gibbs, R., 1967, Amazon River: environmental factors that control its dissolved and suspended load: Science, v. 156, 1734-1737.

* Gibbs, R., 1973, Mechanisms of trace metal transport in rivers: Science, v. 180, p. 71-73

* Gibbs, R., 1977, Transport phases of transition metals in the Amazon and Yukon Rivers: Geological Society of America Bulletin, v. 88, p. 829-843.

Gibbs, R., 1983, Effect of natural organic coatings on the coagulation of particles: Environmental Science and Technology, v. 17, p. 237-240.

* Goldberg, E., 1954, Marine geochemistry I - chemical scavengers of the sea: Journal of Geology, v. 62, p. 249-265.

* Goldberg, E., and Arrhenius, G., 1958, Chemistry of Pacific pelagic sediments: Geochimica et Cosmochimica Acta, v., 13, p. 153-212.

* Green, R., 1979, Sampling design and statistical methods for environmental biologists, New York, John Wiley & Sons, 257 p.

* Gruebel, K., Davis, J., and Leckie, J., 1988, The feasibility of using sequential extraction techniques for arsenic and selenium in soils and sediments: Soil Science Society of America Journal, v. 52, p. 390-397.

* Grim, R., 1968, Clay mineralogy, 2nd edition, New York, McGraw-Hill, 596 p.

Gunn, A., Hunt, D., and Winnard, D., 1989, The effect of heavy metal speciation in sediment on bioavailability to tubificid worms: Hydrobiologia, v. 188/189, p. 487-496.

* Gupta, S., and Chen, K., 1975, Partitioning of trace metals in selective chemical fractions of near-shore sediments: Environmental Letters, v. 10, p. 129-158.

* Guy, H., 1966, System for monitoring fluvial sediment , in, Selected techniques for water resources investigations, U.S. Geological Survey Water Supply Paper 1822, p. 84-88.

* Guy, H., 1969, Laboratory theory and methods for sediment analysis: U.S. Geological Survey Techniques of Water Resources Investigations, book 5, chapter C1, 58 p.

* Gy, P., 1979, Sampling of particulate materials: theory and practice, New York, Elsevier, 431 p.

* Hakanson, L., 1984, Sediment sampling in different aquatic environments - statistical aspects: Water Resources Research, v. 20, p. 41-46.

Hart, B., 1982, Uptake of trace metalsby sediments and suspended particulates: a review: Hydrobiologia, v. 91, p. 299-313.

* Hawkes, H., and Webb, J., 1962, Geochemistry in mineral exploration, New York, Harper and Row, 415 p.

Heath, G., and Dymond, J., 1977, Genesis and transformation of metalliferous sediments from the East Pacific Rise, Bauer Deep, and Central Basin, northwest Nazca Plate: Geological Society of America Bulletin, v. 88, p. 723-733.

* Heller-Kellai, L., and Rozenson, I., 1981, Nontronite after acid or alkali attack: Chemical Geology, v. 32, p. 95-102.

Hem, J., and Robertson, C., 1967, Form and stability of aluminum hydroxide complexes in dilute solution: U.S. Geological Survey Water Supply Paper, 1827-A, p. A24.

Hem, J., and Robertson, C., Lind, C., and Polzer, W., 1973, Chemical interactions of aluminum with aqueous silica at 25°C: U.S. Geological Survey Water Supply Paper, 1827-E.

* Hirner, A., Kritsotakis, K., and Tobschall, H., 1990, Metal-organic associations in sediments - I, comparison of unpolluted recent and ancient sediments and sediments affected by anthropogenic pollution: Applied Geochemistry, v. 5, p. 491-506.

* Hirst, D., 1962, The geochemistry of modern sediments from the Gulf of Paria: Geochimica et Cosmochimica Acta, v. 26, p. 1147-1187.

* Hirst, D., and Nicholls, G., 1958, Techniques in sedimentary geochemistry, I, separation of the detrital and non-detrital fractions of limestones: Journal of Sedimentary Petrology, v. 28, p. 468-481.

* Hjulstrom, F., 1935, Studies of the morphological activity of rivers as illustrated by the River Fyris: Bulletin of the Geological Institute of Upsala, Upsala, Sweden, v. XXV.

Holmgren, G., 1967, A rapid citrate-dithionate extractable iron procedure: Soil Science Society of America Proceedings, v. 31, p. 210-211.

* Horowitz, A., 1974, The geochemistry of sediments from the northern Reykjanes Ridge and the Iceland-Faroes Ridge: Marine Geology, v. 17, p. 103-122.

* Horowitz, A., 1986, Comparison of methods for the concentration of suspended sediment in river water for subsequent chemical analysis: Environmental Science and Technology, v. 20, p. 155-160.

* Horowitz, A., 1990, The role of sediment-trace element chemistry in water-quality monitoring and the need for standard analytical methods, in, Hall, J., and Glysson, D., eds., Monitoring water in the 1990's: meeting new challenges, ASTM STP 1102, Philadelphia, American Society for Testing and Materials, (in press).

* Horowitz, A., and Cronan, C., 1976, The geochemistry of basal sediments from the North Atlantic Ocean: Marine Geology, v. 20, p. 205-228.

* Horowitz, A., and Elrick, K., 1985, Multielement analysis of rocks and sediments by wet digestion and atomic absorption spectroscopy, Varian Instruments at Work, AA-47, 7p.

* Horowitz, A., and Elrick, K., 1986, An evaluation of air elutriation for sediment particle size separation and subsequent chemical analysis: Environmental Technology Letters, v. 7, p. 17-26.

* Horowitz, A., and Elrick, K., 1987, The relation of stream sediment surface area, grain size, and composition to trace element chemistry: Applied Geochemistry, v. 2, p. 437-451.

* Horowitz, A., and Elrick, K, 1988, Interpretation of bed sediment trace metal data: methods for dealing with the grain size effect, in Lichtenberg, J., and others, eds., Chemical and biological characterization of sludges, sediments, dredge spoils, and drilling muds, ASTM STP 976, Philadelphia, Penn., p. 114-128.

* Horowitz, A., Elrick, K., and Callender, E., 1988, The effect of mining on the sediment-trace element geochemistry of cores from the Cheyenne River arm of Lake Oahe, South Dakota, USA: Chemical Geology, v. 67, p. 17-33.

* Horowitz, A., Elrick, K., and Cook, R., 1989d, Source and transport of arsenic in the Whitewood Creek-Belle Fourche-Cheyenne River-Lake Oahe system, South Dakota, in, Mallard, G., and Ragone, S., eds., U.S. Geological Survey toxic substances hydrology program - proceedings of the technical meeting, Phoenix, AZ, Sept. 26-30, 1988, U.S. Geological Survey Water Resources Investigation Report 88-4220, p. 485-493.

* Horowitz, A., Elrick, K., and Hooper, R., 1989a, The prediction of aquatic sediment-associated trace element concentrations using selected geochemical factors: Hydrological Processes, v. 3, p. 347-364.

* Horowitz, A., Elrick, K., and Hooper, R., 1989b, A comparison of instrumental dewatering methods for the separation and concentration of suspended sediment for subsequent trace element analysis: Hydrological Processes, v. 2, p. 163-184.

* Horowitz, A., Rinella, F., Lamothe, P., Miller, T., Edwards, T., Roche, R., and Rickert, D., 1989c, Cross-sectional variability in suspended sediment and associated trace element concentrations in selected rivers in the U.S., in, Hadley, R., and Ongley, E., eds., Sediment and the environment, IAHS Publication No. 184, p. 57-66.

* Horowitz, A., Rinella, F., Lamothe, P., Miller, T., Edwards, T., Roche, R., and Rickert, D., 1990, Variations in suspended sediment and associated trace element concentrations in selected riverine cross sections: Environmental Science and Technology, v. 24, 1313-1320.

* Irving, H., and Williams, R., 1948, Order of stability of metal complexes: Nature (London), v. 162, p. 746-747.

Jackson, M., 1958, Soil chemical analysis: Englewood Cliffs, N.J., Prentice-Hall, 498 p.

* Jackson, M., 1979, Soil chemical analysis - advanced course, 2nd edition, Madison, Wis., published by the author, 898 p.

Jean, G., and Bancroft, G., 1986, Heavy metal adsorption by sulphide mineral surfaces: Geochmica et Cosmochimica Acta, v. 50, p. 1455-1463.

* Jenne, E., 1968, Controls of Mn, Fe, Co, Ni, Cu, and Zn concentrations in soils and water: the significance of hydrous Mn and Fe oxides: Advances in Chemistry Series, v. 73, p. 337-387.

* Jenne, E., 1976, Trace element sorption by sediments and soils - sites and processes, in Chappell, W., and Peterson, K., eds., Symposium on molybdenum, v. 2, New York, Marcel-Dekker, p. 425-553.

Jenne, E., 1981, Speciation of aqueous contaminants - role of the geochemical model, in Proceedings of the NBS Workshop on Aqueous Speciation of Dissolved Contaminants, Gaithersburg, Maryland, CONF 810588-4, NTIS, 14 p.

* Jenne, E., Kennedy, V., Burchard, J., and Ball, J., 1980, Sediment collection and processing for selective extraction and for total metal analysis, in Baker, R., ed., Contaminants and sediments, v. 2, Ann Arbor, Mich., Ann Arbor Science Publishers, Inc., p. 169-189.

Jenne, E., and Luoma, S., 1977, Forms of trace elements in soild, sediments, and associated waters: an overview of their determination and biological availability, in Wildung, R., and Drucker, H., eds., Biological implications of metals in the environment, CONF-750929, NTIS, Springfield, Va, p. 110-143.

* Johnson, W., and Maxwell, J., 1981, Rock and mineral analysis, 2nd edition, New York, John Wiley, 489 p.

* Jonasson, I., 1977, Geochemistry of sediment/water interactions of metals, including observations on availability, in Shear, H., and Watson, A., eds., The fluvial transport of sediment-associated nutrients and contaminants, IJC/PLUARG, Windsor, Ontario, p. 255-271.

* Jones, B., and Bowser, C., 1978, The mineralogy and related chemistry of lake sediments, in Lerman, A., ed., Lakes: chemistry, geology, physics, New York, Springer-Verlag, p. 179-235.

* Kaurichev, I., Vityazev, V., and Shevchenko, A., 1983, Effects of non-silicate forms of iron and humus on specific soil surfaces: Izvestia Timiryazevskoi Sel'skokhozyaistvennoi Akademii, Number 1, p. 103-106.

* Keith, L.,1988, Principles of environmental sampling, American Chemical Society, Washington, D.C., 458 p.

* Keith, L., Crummett, W., Deegan, J., Libby, R., Taylor, J., and Wentler, G., 1983, Principles of environmental analysis: Analytical Chemistry, v. 55, p. 2210-2218.

Kempton, S., Sterritt, R., and Lester, J., 1987, Heavy metal removal in primary sedimentation II. the influence of metal speciation and particle size distribution: The Science of the Total Environment, v. 63, p. 247-258.

Kennedy, V., Zellwager, G., and Jones, B., 1974, Filter pore-size effects on the analysis of Al, Fe, Mn, and Ti in water: Water Resources Research, v. 10, p. 785-790.

Keulder, P., 1982, Particle size distribution and chemical parameters of the sediments of a shallow turbid impoundment: Hydrobiologia, v., 91, p. 341-353.

* Kharkar, D., Turekian, K., and Bertine, K., 1968, Stream supply of dissolved Ag, Mo, Sb, Se, Cr, Ca, Rb, and Cs to the oceans: Geochimica et Cosmochimica Acta, v. 33, p. 285-298.

* Kheboian, C., and Bauer, C., 1987, Accuracy of selective extraction procedures for metal speciation in model aquatic sediments: Analytical Chemistry, v. 59, p. 1417-1423.

Kitano, Y., Sakata, M., and Matsumoto, E., 1980, Partitioning of heavy metals into mineral and organic fractions in a sediment core from Tokyo Bay: Geochimica et Cosmochimica Acta, v. 22, p. 1279-1285.

Klock, P., Czamanske, G., Foose, M., and Pesek, J., 1986: Chemical Geology, v. 54, p. 157-163.

* Kononova, M., 1966, Soil organic matter, 2nd edition, Nowakowski, T., and Newman, A., translators, New York, Pergamon Press, p. 377-419.

* Krauskopf, K., 1956, Factors controlling the concentration of thirteen rare metals in sea water: Geochimica et Cosmochimica Acta, v. 9, p. 1-32.

* Krumbein, W., and Pettijohn, F., 1938, Manual of sedimentary petrography, New York, Appleton-Century-Crofts, Inc., 549 p.

* Kuenen, Ph., 1965, Geological conditions of sedimentation, in Riley, J., and Skirrow, G., eds., Chemical oceanography, v.2, New York, Academic Press, p. 1-22.

Laxen, D., 1985a, Adsorption of Cd, Pb, and Cu during the precipitation of hydrous ferric oxide in a natural water: Chemical Geology, v. 47, p. 321-332.

Laxen, D., 1985b, Trace metal adsorption/coprecipitation on hydrous ferric oxide under realistic conditions, the role of humic substances: Water Research, v. 19, p. 1229-1236.

Laxen, D., and Chandler, I., 1983, Size distribution of iron and manganese species in freshwaters: Geochimica et Cosmochimica Acta, v. 47, p. 731-741.

Leckie, J., 1986, Adsorption and transformation of trace element species at sediment/water interfaces, in Bernard, M., and others, eds., The importance of chemical speciation in environmental processes, Dahlem Workshop Report, v. 33, Berlin, Springer-Verlag, p. 237-254.

* Lee, G., 1975, Role of hydrous metal oxides in the transport of heavy metals in the environment, in Krenkel, P., ed., Heavy metals in the aquatic environment, New York, Pergamon Press, p. 137-147.

* Leinen, M., and Pisias, N., 1984, An objective technique for determining end member compositions and for partitioning sediments according to their sources: Geochimica et Cosmochimica Acta, v. 48, p. 47-62.

* Leinen, M., and Pisias, N.,Dymond, J., and Heath, G., 1980, Geochemical partitioning: application of an objective technique for end-member characterization: Geological Society of America Abstracts with Programs, v. 93, p. 470 (abstract).

Lewis, T., and McIntosh, A., 1989, Covariance of selected trace elements with binding substrates in cores collected from two contaminated sediments: Bulletin of Environmental Contamination and Toxicology, v. 43, p. 518-528.

* Lichtenberg, J., Winter, J., Weber, C., and Fradkin, L., 1988, Chemical and biological characterization of municipal sludges, sediments, dredge spoils, and drilling muds, ASTM STP 976, Philadelphia, PA, American Society for Testing and Materials, 512 p.

Lion, L., Altmann, R., and Leckie, J., 1982, Trace metal adsorption characteristics of estuarine particulate matter: evaluation of contributions of Fe/Mn oxide and organic surface coatings: Environmental Science and Technology, v. 16, p. 660-666.

Liu, B., Raabe, O., Smith, W., Spencer, H., III, and Kuykendal, W., 1980, Advances in particle sampling and measurement: Environmental Science and Technology, v. 14, p. 392-397.

Loring, D., 1981, Potential bioavailability of metals in eastern Canadian estuarine and coastal sediments: Rapports et Proces-Verbaux des Reunions Conseil International Pour l'Exploration de la Mer, v. 181, p. 93-101.

* Loring, D., 1990, Lithium - a new approach for the granulometric normalization of trace metal data: Marine Chemistry, v. 29, p. 155-168.

* Love, S., 1960, Quality of surface waters of the United States, 1958, U.S. Geological Survey Water Supply Paper 1571, Parts 1-4, 733 p.

* Lovell, V., 1975, The effect of certain pretreatments on the surface area of various minerals determined by the BET method: Powder Technology, v. 12, p. 71-76.

Luoma, S., 1983, Bioavailability of trace metals to aquatic organisms - a review: The Science of the Total Environment, v. 28, p. 1-22.

Luoma, S., 1986, A comparison of two methods for determining copper partitioning in oxidized sediments: Marine Chemistry, v. 20, 45-59.

Luoma, S., 1989, Can we determine the biological availability of sediment-bound trace elements?: Hydrobiologia, v. 176/177, p. 379-396.

Luoma, S., and Bryan, G., 1978, Factors controlling the availability of sediment-bound lead to the estuarine bivalve *Scrobicularia plana*: Journal of the Marine Biological Association of the United Kingdom, v. 58, p. 793-802.

Luoma, S., and Bryan, G., 1979a, Trace metal bioavailability: modelling chemical and biological interactions of sediment-bound zinc, in Jenne, E., ed., Chemical modelling in aqueous systems, ACS Symposium Series 93, American Chemical Society, Washington, D.C., p. 577-609.

* Luoma, S., and Bryan, G., 1981, A statistical assessment of the form of trace metals in oxidized estuarine sediments employing chemical extractants: The Science of the Total Environment, v. 17, p. 165-196.

* Luoma, S., and Davis, J., 1983, Requirements for modelling trace metal partitioning in oxidized estuarine sediments: Marine Chemistry, v. 12, p. 159-181.

Luoma, S., and Jenne, E., 1976, Factors affecting the availability of sediment-bound cadmium to the estuarine deposit-feeding clam, *Macoma baltica*, in Radiological problems associated with the development of energy sources, Proceedings of the Fourth National Symposium on Radioecology, Corvallis, Oreg., ERDA Report CONF-759503, Stroudsburg, Pa., Dowden, Hutchinson and Russ Publishers, 16 p.

* Luoma, S., and Jenne, E., 1977a, Estimating bioavailability of sediment-bound metals with chemical extractants, in Hemphill, D., ed., Trace substances in environmental health, v. 10, Columbia, Mo., University of Missouri Press, p. 343-351.

Luoma, S., and Jenne, E., 1977b, The availability of sediment-bound cobalt, silver, and zinc to a deposit-feeding clam, in Wildung, R., and Drucker, H., eds., Biological implications of metals in the environment, CONF-750929, NTIS, Springfield, Va., p. 213-230.

* Lynn, D., and Bonatti, E., 1965, Mobility of manganese in diagenesis of deep sea sediments: Marine Geology, v. 3, p. 457-474.

* Malo, B., 1977, Partial extraction of metals from aquatic environments: Environmental Science and Technology, v. 11, p. 277-282.

* Martin, J., and Meybeck, M., 1979, Elemental mass-balance of material carried by world major rivers: Marine Chemistry, v. 7, p. 173-206.

Martin, H., Wilhelm, E., Laville-Timsit, L., and Lecomte, P., 1984, Enhancement of stream-sediment geochemical anomalies in Belgium and France by selective extractions and mineral separations: Journal of Geochemical Exploration, v. 20, p. 179-205.

* Mero, J., 1962, Occurrence of Mn nodules: Economic Geology, v. 57, p. 747-767.

* Meybeck, M., and Helmer, R., 1989, The quality of rivers: from pristine stage to global pollution: Paleogeography, Paleoclimatology, and Paleoecology (Global and Planetary Change Section), v. 75, p. 283-309.

* Micromeritics, 1986, Instruction manual for the Flowsorb II 2300 for determining single point and multipoint surface area, total pore volume, and pore area and volume distribution, Norcross, GA, Micromeritics Corp., p. A-12 - A-14.

Miller, W., Martens, D., and Zelazny, L, 1986, Effect of sequence in extraction of trace metals from soils: Soil Science Society of America Journal, v. 50, p. 598-601.

* Mitchell, R., 1964, Trace elements in soil, in Bear, F., ed., Chemistry of the soil, New York, Reinhold Publishing Corp., p. 320-368.

* Moore, R., Meyer, R., and Morgan, C., 1973, Investigation of the sediments and potential manganese nodule resources of Green Bay, Wisconsin: University of Wisconsin Technical Report WIS-SG-73-213, Madison, Wis., 144 p.

Mora, S., and Harrison, R., 1983, The use of physical separation techniques in trace metal speciation studies: Water Research, v. 17, p. 723-733.

Mouvet, C., and Bourg, A., 1983, Speciation (including adsorbed species) of copper, lead, nickel, and zinc in the Meuse River: Water Research, v. 17, p. 641-649.

* Muller, L, and Burton, C., 1965, The heavy liquid density gradient and its application in ore dressing mineralogy: Proceedings of the eighth commonwealth mining and metallurgical congress, v. 6, Melbourne, Australia, p. 1151-1163.

Musani, Lj., Valenta, P., Nurnberg, H., Konrad, Z., and Brancia, M., 1980, On the chelation of toxic trace metals by humic acid of marine origin: Estuarine and Coastal Marine Research, v. 11, p. 639-649.

* National Institute of Standards and Technology (NIST), 1990, NIST standard reference materials catalog 1990-1991, Gaithersburg, MD, U.S. Department of Commerce, National Institute of Standards and Technology, 161 p.

Nembrini, G., Rapin, F., Garcia, J., and Forstner, U., 1982, Speciation of Fe and Mn in a sediment core of the Baie de Villefrance: Environmental Technology Letters, v.3, p. 545-552.

Nishida, H., Mayai, M., Tada, F., and Suzuki, S., 1982, Computation of the index of pollution caused by heavy metals in river sediment: Environmental Pollution (Series B), v. 4, p. 241-248.

Nissenbaum, A., 1972, Distribution of several metals in chemical fractions of a sediment core from the Sea of Okhotsk: Israeli Journal of Earth Science, v. 21, p. 143-154.

* Norris, J., 1988, Techniques for sampling surface and industrial waters: special considerations and choices, in, Keith, L., ed., Principles of environmental sampling, Washington, D.C., American Chemical Society, p. 247-253.

* Nriagu, J., and Coker, R., 1980, Trace metals in humic and fulvic acids from Lake Ontario sediments: Environmental Science and Technology, v. 14, p. 443-446.

Nriagu, J., Wong, H., and Snodgrass, W., 1983, Historical record of metal pollution in sediments of Toronto and Hamilton harbours: Journal of Great Lakes Research, v. 9, p. 365-373.

* Oakley, S., Nelson, P., and Williamson, K., 1981, Model of trace-metal partitioning in marine sediments: Environmental Science and Technology, v., 15, p. 474-480

* Oakley, S., Williamson, K., and Nelson, P., 1980, The geochemical partitioning and bioavailability of trace metals in marine sediments: Project completion report, OWRR Project No. A-044-ORE-WF3, October 1, 1977-September 31, 1979, Oregon State University, Corvallis, Oreg., 84 p.

* Oliver, B., 1973, Heavy metal levels of Ottawa and Rideau River sediment: Environmental Science and Technology, v. 7, p. 135-137.

Olsen, C., Cutshall, N., and Larsen, I., 1982, Pollutant-particle associations and dynamics in coastal marine environments: a review: Marine Chemistry, v. 11, p. 501-533.

* Ongley, E., Birkholz, D., Carey, J., and Samoiloff, M., 1988, Is water a relevant sampling medium for toxic chemicals? an alternative environmental sensing strategy: Journal of Environmental Quality, v. 17, p. 391-401.

* Ongley, E., and Blachford, N., 1982, Application of continuous-flow centrifugation to contaminant analysis of suspended sediment in fluvial systems: Environmental Technology Letters, v. 3, p. 219-228.

* Office of Water Data Coordination (OWDC), 1982, National handbook of recommended methods for water-data acquisition, U.S. Geological Survey, Reston, VA, Chapter3 - Sediment, p. 3-1 - 3-100.

Patchineelam, S., 1975, Untersuchungen uber die hauptbindungsarten und die mobilisierburkeit von schwermetallen in fluviatilen sedimenten: unpublished Ph.D. dissertation, University of Heidelberg, 136 p.

Patchineelam, S., and Forstner, U., 1977, Bildungsformen von schwermetallen in mariene sedimenten: Senckenbergiana Maritima, v. 9, p. 75-104.

Pavoni, B., Marcomini, A., Sfriso, A., Orio, A.A., 1988, Multivariate analysis of heavy metal concentrations in sediments of the lagoon of venice: The Science of the Total Environment, v. 77, p. 189-202.

Pickering, W., 1981, Selective chemical extraction of soil components and bound metal species: Critical Reviews of Analytical Chemistry, v. 12, p. 233-266.

* Pilkington, E., and Warren, L., 1979, Determination of heavy metal distribution in marine sediments: Environmental Science and Technology, v. 13, p. 295-299.

Piper, D., 1971, The distribution of Co, Cr, Cu, Fe, Mn, Ni, and Zn in Framvaren, a Norwegian anoxic fjord: Geochimica et Cosmochimica Acta, v. 35, p. 531-550.

* Piper, D., 1973, Origin of metalliferous sediments from the East Pacific Rise: Earth and Planetary Science Letters, v. 19, p. 75-82.

* Pisias, N., and Leinen, M., 1980, Geochemical paritioning of deep-sea sediments using an extended version of Q-mode factor analysis and linear programming: Geological Society of America Abstracts with Programs, v. 93, p. 500 (abstract).

* Plumb, R., Jr., 1981, Procedures for handling and chemical analysis of sediment and water samples, Technical Report EPA/CE-81-1, Environmental Laboratory, U.S. Army Engineer Waterways Experiment Station, Vicksburg, p. 3-73 - 3-76.

* Porterfield, G., 1972, Computation of fluvial sediment discharge: U.S. Geological Survey Techniques of Water Resources Investigations, book 3, chapter C-3, p. 43-47.

Rapin, F., Tessier, A., Campbell, P., and Carignan, R., 1986, Potential artifacts in the determination of metal partitioning in sediments by a sequential extraction procedure: Environmental Science and Technology, v. 20, p. 836-840.

* Rashid, M., 1974, Adsorption of metals on sedimentary and peat humic acids: Chemical Geology, v., 13, p. 115-123.

* Raudkivi, A., 1967, Analyses of resistance in fluvial channels: Journal of the Hydraulics Division, American Society of Civil Engineers, v. 93, No. HY5, Proceedings Paper 5426, September 1967, p. 73-84.

Renberg, I., 1986, Concentration and annual accumulation values of heavy metals in lake sediments: their significance in studies of the history of heavy metal pollution: Hydrobiologia, v. 143, p. 379-385.

* Renzoni, A., Bacci, E., and Falcia, L., 1973, Mercury concentration in the water, sediments, and fauna of an area of the Tyrrhenian coast, Revue Internationale d'Oceanographie Medicale, v. 31/32, p. 17-45.

* Rex, R., and Goldberg, E., 1958, Quartz contents of pelagic sediments of the Pacific Ocean: Tellus, v. 10, p. 153-159.

* Robbins, J., Lyle, M., and Heath, G., 1984, A sequential extraction procedure for partitioning elements among co-existing phases in marine sediments, Contribution 84-3, College of Oceanography, Oregon State University, Corvallis, Oreg., 64 p.

Robinson, G., 1981, Adsorption of Cu, Zn, and Pb near silfide deposits by hydrous manganese-iron oxide coatings on stream alluvium: Chemical Geology, v. 12, p. 233-266.

* Robinson, G., 1982, Trace metal adsorption potential of phases comprising black coatings on stream pebbles: Journal of Geochemical Exploration, v. 17, p. 205-219.

Robinson, G., 1983, Heavy-metal adsorption by ferromanganese coatings on stream alluvium: natural controls and implications for exploration: Chemical Geology, v. 38, p. 157-174.

Robinson, G., 1985, Sequential chemical extractions and metal partitioning in hydrous Mn-Fe oxide cotaings: reagent choice and substrate composition affect results: Chemical Geology, v. 47, p. 97-112.

* Rossman, R., Callender, E., and Bowser, C., 1972, Interelement geochemistry of Lake Michigan ferromanganese nodules: Proceedings of the 24th international geological congress, Montreal, Section 10, p. 336-341.

Saar, R., and Weber, J., 1981, Lead (II) complexation by fulvic acid: how it differs from fulvic acid complexation of copper (II) and cadmium (II): Geochimica et Cosmochimica Acta, v. 44, p. 1381-1384.

Salomons, W., 1985, Sediments and water quality: Environmental Technology Letters, v. 6, p. 315-326.

* Salomons, W., and Forstner, U., 1984, Metals in the hydrocycle, New York, Springer-Verlag. 349 p.

* Sanders, T., Ward, R., Loftis, J., Steele, T., Adrian, D., and Yevjevich, V., 1983, Design of networks for monitoring water quality, Littleton, CO, Water Resources Publications, 328 p.

* Saxby, J., 1969, Metal-organic chemistry of the geochemical cycle: Reviews of Pure and Applied Chemistry, v. 19, p. 131-150.

Schalscha, E., Morales, M., Vergara, I., and Chang, A., 1982, Chemical fractionation of heavy metals in wastewater affected soils: Journal of the Water Pollution Control Federation, v. 54, p. 175-180.

* Schmidt, R., Garland, T., and Wildung, R., 1975, Copper in Sequim Bay sediments: Batelle Pacific Northwest Laboratory Annual Report, part 2, p. 136.

Schnitzer, M., and Kerndorff, H., 1981, Reactions of fulvic acid with metal ions: Water, Air, and Soil Pollution, v. 15, p. 97-108.

* Schnitzer, M., and Kahn, S., 1972, Humic substances in the environment, New York, Marcel-Dekker, 327 p.

Schnitzer, M., and Kahn, S., 1978, Soil organic matter, New York, Elsevier, p. 1-64.

Schultz, L., 1964, Quantitative interpretation of mineralogical composition from x-ray and chemical data for the Pierre Shale: U.S. Geological Survey Professional Paper 391-C, 31 p.

Schwertmann, U., 1964, Differenzierung der eisenoxide des bodens photochemische extraktion mit sauer ammonium-oxalat-losung: Zeitschrift fur Pflanzenerahrung, Dungung mit Bodenkunde, v. 105, p. 194-202.

* Singer, P., 1977, Influence of dissolved organics on the distribution, transport, and fate of heavy metals in aquatic systems, in Suffet, I., ed., Fate of pollutants in the air and water environment, part I, New York, John Wiley, p. 155-182.

Singh, S., and Subramian, V., 1983, Hydrous Fe and Mn oxides - scavengers of heavy metals in the aquatic environment: CRC Critical Reviews in Environmental Control, v. 14, p. 33-90.

Skei, J., and Paus, P., 1979, Surface metal enrichment and partitioning of metals in a dated sediment core from a Norwegian fjord: Geochimica et Cosmochimica Acta, v. 43, p. 239-246.

Skoog, D., and West, D., 1981, Principles of instrumental analysis, 2nd edition, Philadelphia, PA, Saunders College Press, p. 452-457.

* Skougstad, M., Fishman, M., Friedman, L., Erdmann, D., and Duncan, S., 1979, Methods for determination of inorganic substances in water and fluvial sediments, U.S. Geological Survey Techniques of Water Resources Investigations, book 5, Chapter A1, p. 561.

Slavek, J., and Pickering, W., 1981, The effect of pH on the retention of Cu, Pb, Cd, and Zn bt clay-fulvic acid mixtures: Water, Air, and Soil Pollution, v. 16, p. 209-221.

Slavek, J., and Pickering, W., 1985, Chemical leaching of metal ions sorbed on hydrous manganese oxide: Chemical Geology, v. 51, p. 213-223.

* Sly, P., 1969, Bottom sediment sampling, in, Proceedings of the 12th conference on Great Lakes Research, International Association of Great Lakes Research, Ann Arbor, University of Michigan Press, p. 883-898.

Sposito, G., Lund, L., and Chang, A., 1982, Trace metal chemistry in arid-zone field soils phases: Soil Science Society of America Journal, v. 46, p. 260-264.

* Stockham, J., and Fochtman, E., 1979, Particle size analysis, Ann Arbor, Ann Arbor Science Publishers, 140 p.

* Stoffers, P., Summerhayes, C., Forstner, U., and Patchineelem, S., 1977, Copper and other heavy metal contamination in sediments from New Bedford Harbor, Massachusetts: a preliminary note: Environmental Science and Technology, v. 11, p. 819-821.

Stover, R., Sommers, L., and Silviera, D., 1976, Evaluation of metals in wastewater sludge: Journal of the Water Pollution Control Federation, v. 48, p. 2165-2175.

* Suess, E., 1973, Interaction of organic compounds with calcium carbonate-II organo-carbonate association in recent sediments: Geochmica et Cosmochimica Acta, v. 37, p. 2435-2447.

* Summerhayes, C., Ellis, J., Stoffers, P., Briggs, S., and Fitzgerald, M., 1976, Fine-grained sediment and industrial waste distribution in New Bedford Harbor and western Buzzards Bay, Massachusetts: Woods Hole Oceanographic Institution Technical Report WHOI-76-115, 110 p.

* Swallow K., and Morel, F., 1980, Adsorption of trace metals by hydrous ferric oxides in seawater: Environmental Protection Agency Report No. EPA 600/3-80-011, Environmental Research Laboratory, Narragansett, R.I., 52 p.

* Swanson, V., Frist, L., Radar, R., Jr., and Huffman, C., Jr., 1966, Metal sorption by northwest Florida humate: U.S. Geological Survey Professional Paper 550-C, p. 174-177.

* Tessier, A., and Campbell, P., 1987, Partitioning of trace metals in sediments: relationship with bioavailability, in Thomas, R., et al. (eds.), Ecological Effects of *in situ* sediment contaminants: Hydrobiologia, v. 149, 43-52.

* Tessier, A., Campbell, P., and Bisson, M., 1979, Sequential extraction procedure for the speciation of particulate trace metals: Analytical Chemistry, v. 51, p. 844-851.

Tessier, A., Campbell, P., and Bisson, M., 1982, Particulate trace metal speciation in stream sediments and relationships with grain size: implications for geochemical exploration: Journal of Geochemical Exploration, v. 16, p. 77-104.

Tessier, A., Rapin, F., and Carignan, R., 1985, Trace metals in oxic lake sediments: possible adsorption onto iron oxyhydroxides: Geochimica et Comochimica Acta, v. 49, p. 181-194.

* Thomas, R., 1972, The distribution of mercury in the sediment of Lake Ontario: Canadian Journal of Earth Science, v. 9, p. 636-651.

* Thomas, R., Kemp, A., and Lewis, C., 1972, Distribution, composition, and characteristics of the surficial sediments of Lake Ontario: Journal of Sedimentary Petrology, v. 42, 66-82.

Thomson, E., Luoma, S., Johansson, C., and Cain, D., 1984, Comparison of sediments and organisms in identifying sources of biologically available trace metal contamination: Water Research, v. 18, p. 755-765.

* Thorne, L, and Nickless, G., 1981, The relation between heavy metals and particle size fractions within the Severn estuary (U.K.) inter-tidal sediments: The Science of the Total Environment, v. 19, p. 207-213.

* Todd, W., Hogan, J., and Spaite, P., 1963, Test dust preparation and evaluation, Robert A. Taft Sanitary Engineering Center, U.S. Public Health Service, Cincinatti, 9 p.

* U.S. Interagency Committee, 1940, Field practice and equipment used in sampling suspended sediment, St. Paul Engineer District Sub-Office, Hydraulic Laboratory, University of Iowa, Iowa City, Report No. 1, 175 p.

* U.S. Interagency Committee, 1941, Methods of analyzing sediment samples, St. Paul Engineer District Sub-Office, Hydraulic Laboratory, University of Iowa, Iowa City, Report No. 4, 203 p.

van Valin, R., and Morse, J., 1982, An investigation of methods commonly used for the selective removal and characterization of trace metals in sediments: Marine Chemistry, v. 11, p. 553-564.

* Vanoni, V., 1977, Sedimentation engineering, New York, American Society of Civil Engineers, Task Committee for the Preparation of the Manual on Sedimentation of the Hydraulics Division, p. 17-316.

* Vuceta, J., and Morgan, J., 1978, Chemical modelling of trace metals in fresh waters: role of complexation and adsorption: Environmental Science and Technology, v. 12, p. 1302-1309.

Walling, D., and Moorehead, P., 1989, The particle size characteristics of fluvial suspended sediment: an overview: Hydrobiologia, v. 176/177, p. 125-149.

* Watterson, J., and Theobald, P., 1979, Geochemical exploration, 1978, Proceedings of the 7th International Geochemical Exploration Symposium, Toronto, Canada, Association of Exploration Geochemists, 504 p.

* Webb, J., 1978, The Wolfson geochemical atlas of England and Wales, Oxford, Oxford University Press, p. 7-14.

* Webster, D., 1989, A semiquantitative x-ray diffraction method to determine mineral composition in stream sediments with similar mineralogy: Environmental Technology Letters, v. 10, p. 833-844.

Young, T., Delpinto, J., and Seger, E., 1982, Transport and fate of heavy metals in Onondaga Lake, N.Y.: Bulletin of Environmental Contamination and Toxicology, v. 29, p. 554-561.

Ziper, C., Komarneni, S., and Baker, D., 1988, Specific cadmium sorption in relation to the crystal chemistry of clay minerals: Soil Science Society of America Journal, v. 52, p. 49-53.

Index